LONDON MATHEMATICAL SOCIETY LECTURE NOTE SERIES

Managing Editor: Professor I.M. James,
Mathematical Institute, 24-29 St Giles,Oxford

1. General cohomology theory and K-theory, P.HILTON
4. Algebraic topology, J.F.ADAMS
5. Commutative algebra, J.T.KNIGHT
8. Integration and harmonic analysis on compact groups, R.E.EDWARDS
9. Elliptic functions and elliptic curves, P.DU VAL
10. Numerical ranges II, F.F.BONSALL & J.DUNCAN
11. New developments in topology, G.SEGAL (ed.)
12. Symposium on complex analysis, Canterbury, 1973, J.CLUNIE & W.K.HAYMAN (eds.)
13. Combinatorics: Proceedings of the British Combinatorial Conference 1973, T.P.McDONOUGH & V.C.MAVRON (eds.)
15. An introduction to topological groups, P.J.HIGGINS
16. Topics in finite groups, T.M.GAGEN
17. Differential germs and catastrophes, Th.BROCKER & L.LANDER
18. A geometric approach to homology theory, S.BUONCRISTIANO, C.P. ROURKE & B.J.SANDERSON
20. Sheaf theory, B.R.TENNISON
21. Automatic continuity of linear operators, A.M.SINCLAIR
23. Parallelisms of complete designs, P.J.CAMERON
24. The topology of Stiefel manifolds, I.M.JAMES
25. Lie groups and compact groups, J.F.PRICE
26. Transformation groups: Proceedings of the conference in the University of Newcastle-upon-Tyne, August 1976, C.KOSNIOWSKI
27. Skew field constructions, P.M.COHN
28. Brownian motion, Hardy spaces and bounded mean oscillations, K.E.PETERSEN
29. Pontryagin duality and the structure of locally compact Abelian groups, S.A.MORRIS
30. Interaction models, N.L.BIGGS
31. Continuous crossed products and type III von Neumann algebras, A.VAN DAELE
32. Uniform algebras and Jensen measures, T.W.GAMELIN
33. Permutation groups and combinatorial structures, N.L.BIGGS & A.T.WHITE
34. Representation theory of Lie groups, M.F. ATIYAH et al.
35. Trace ideals and their applications, B.SIMON
36. Homological group theory, C.T.C.WALL (ed.)
37. Partially ordered rings and semi-algebraic geometry, G.W.BRUMFIEL
38. Surveys in combinatorics, B.BOLLOBAS (ed.)
39. Affine sets and affine groups, D.G.NORTHCOTT
40. Introduction to Hp spaces, P.J.KOOSIS
41. Theory and applications of Hopf bifurcation, B.D.HASSARD, N.D.KAZARINOFF & Y-H.WAN
42. Topics in the theory of group presentations, D.L.JOHNSON
43. Graphs, codes and designs, P.J.CAMERON & J.H.VAN LINT
44. Z/2-homotopy theory, M.C.CRABB
45. Recursion theory: its generalisations and applications, F.R.DRAKE & S.S.WAINER (eds.)
46. p-adic analysis: a short course on recent work, N.KOBLITZ
47. Coding the Universe, A.BELLER, R.JENSEN & P.WELCH
48. Low-dimensional topology, R.BROWN & T.L.THICKSTUN (eds.)

49. Finite geometries and designs, P.CAMERON, J.W.P.HIRSCHFELD & D.R.HUGHES (eds.)
50. Commutator calculus and groups of homotopy classes, H.J.BAUES
51. Synthetic differential geometry, A.KOCK
52. Combinatorics, H.N.V.TEMPERLEY (ed.)
53. Singularity theory, V.I.ARNOLD
54. Markov processes and related problems of analysis, E.B.DYNKIN
55. Ordered permutation groups, A.M.W.GLASS
56. Journées arithmétiques 1980, J.V.ARMITAGE (ed.)
57. Techniques of geometric topology, R.A.FENN
58. Singularities of smooth functions and maps, J.MARTINET
59. Applicable differential geometry, M.CRAMPIN & F.A.E.PIRANI
60. Integrable systems, S.P.NOVIKOV et al.
61. The core model, A.DODD
62. Economics for mathematicians, J.W.S.CASSELS
63. Continuous semigroups in Banach algebras, A.M.SINCLAIR
64. Basic concepts of enriched category theory, G.M.KELLY
65. Several complex variables and complex manifolds I, M.J.FIELD
66. Several complex variables and complex manifolds II, M.J.FIELD
67. Classification problems in ergodic theory, W.PARRY & S.TUNCEL
68. Complex algebraic surfaces, A.BEAUVILLE
69. Representation theory, I.M.GELFAND et al.
70. Stochastic differential equations on manifolds, K.D.ELWORTHY
71. Groups - St Andrews 1981, C.M.CAMPBELL & E.F.ROBERTSON (eds.)
72. Commutative algebra: Durham 1981, R.Y.SHARP (ed.)
73. Riemann surfaces: a view towards several complex variables, A.T.HUCKLEBERRY
74. Symmetric designs: an algebraic approach, E.S.LANDER
75. New geometric splittings of classical knots (algebraic knots), L.SIEBENMANN & F.BONAHON
76. Linear differential operators, H.O.CORDES
77. Isolated singular points on complete intersections, E.J.N.LOOIJENGA
78. A primer on Riemann surfaces, A.F.BEARDON
79. Probability, statistics and analysis, J.F.C.KINGMAN & G.E.H.REUTER (eds.)
80. Introduction to the representation theory of compact and locally compact groups, A.ROBERT
81. Skew fields, P.K.DRAXL
82. Surveys in combinatorics: Invited papers for the ninth British Combinatorial Conference 1983, E.K.LLOYD (ed.)
83. Homogeneous structures on Riemannian manifolds, F.TRICERRI & L.VANHECKE
84. Finite group algebras and their modules, P.LANDROCK
85. Solitons, P.G.DRAZIN
86. Topological topics, I.M.JAMES (ed.)
87. Surveys in set theory, A.R.D.MATHIAS (ed.)

London Mathematical Society Lecture Note Series: 82

Surveys in Combinatorics

Invited papers for the ninth
British Combinatorial Conference: 1983

Edited by E. KEITH LLOYD
Lecturer in the Faculty of Mathematical Studies
University of Southampton

CAMBRIDGE UNIVERSITY PRESS
Cambridge
London New York New Rochelle
Melbourne Sydney

CAMBRIDGE UNIVERSITY PRESS
Cambridge, New York, Melbourne, Madrid, Cape Town,
Singapore, São Paulo, Delhi, Tokyo, Mexico City

Cambridge University Press
The Edinburgh Building, Cambridge CB2 8RU, UK

Published in the United States of America by Cambridge University Press, New York

www.cambridge.org
Information on this title: www.cambridge.org/9780521275521

© Cambridge University Press 1983

This publication is in copyright. Subject to statutory exception
and to the provisions of relevant collective licensing agreements,
no reproduction of any part may take place without the written
permission of Cambridge University Press.

First published 1983
Re-issued 2011

A catalogue record for this publication is available from the British Library

Library of Congress Catalogue Card Number: 83-10078

ISBN 978-0-521-27552-1 Paperback

Cambridge University Press has no responsibility for the persistence or
accuracy of URLs for external or third-party internet websites referred to in
this publication, and does not guarantee that any content on such websites is,
or will remain, accurate or appropriate.

CONTENTS

R.P. STANLEY

V.T. SOS

PREFACE

> For there is no man can Write ſo warily, but that he may ſometime give Opportunity of Cavilling, to thoſe who ſeek it.
>
> John Wallis, A treatiſe of Algebra, 1685.

From rather modest beginnings the British Combinatorial Conference has grown into an established biennial international gathering. A successful format for the Conference has been established whereby several distinguished mathematicians are each invited to give a survey lecture at the Conference and to write a paper for the Conference Volume, which is published in time for the start of the meeting. The present volume contains eight of the nine invited papers for the Ninth Conference held at the University of Southampton, 11-15 July 1983.

Between them the papers cover a broad range of combinatorics. The all-pervading subject of graph theory appears in a number of the papers. It is the central feature of the one by J.C. BERMOND and his co-authors in which they survey those results concerning diameter and connectivity in graphs and hypergraphs of importance for interconnection networks. Graph theory is also used by J.M. HAMMERSLEY in his study of the Friendship Theorem and the Love Problem. His paper looks back to classical mythology with references to Narcissus, but in producing it he has made use of the latest technology in the form of the Oxford University Lasercomp typesetting facility. Perhaps the day is not far off when it will become routine for authors to produce their papers by such means. Other papers using graph theory are those of Schrijver and Shult mentioned below.

J.W.P. HIRSCHFELD writes on Maximal Sets in Finite Projective Spaces, a topic with applications in statistics and coding theory. He has gathered together a huge number of equations and inequalities from a wide literature.

For half a century research on permanents was overshadowed by the van der Waerden conjecture. Recently two proofs of it were obtained independently by Falikman and Egorychev for which work they were jointly awarded one of the Fulkerson Prizes in Discrete Mathematics in August 1982. The paper by A. SCHRIJVER surveying this and related work is, therefore, especially timely. A very different problem which has also been solved recently is that of characterizing the Lie incidence geometries, which are associated with finite simple groups. The solution involves the patching together of results of many researchers and E.E. SHULT'S paper should help to make the subject more accessible.

J. Howard Redfield was a polymath who earned his living as an engineer. From an early age he showed exceptional linguistic ability, later becoming familiar with most European languages as well as some Asian and African tongues. When he turned his attention to combinatorics he produced a paper (published 1927) the value of which was not generally to be recognised by professional combinatorialists until some thirty years later. The recent discovery of a second (unpublished) paper by Redfield shows that he had continued his pioneering work and had obtained further results on enumeration much earlier than other writers. J. SHEEHAN looks at the material in the two Redfield papers and its relation to the work of others including Pólya, Read and himself. Redfield's first paper had opened with the very perceptive statement: "In view of the similarity which will be admitted to hold between the subject matters of the Theory of Finite Groups and of Combinatory Analysis, it is somewhat surprising to find that in their literatures the two branches have proceeded on their separate ways without developing their interrelationship,...". Half a century later this similarity is now well recognized and also extends to certain infinite groups some of which are studied by R.P. STANLEY in his paper on $GL(n,\mathbb{C})$ for Combinatorialists. Nowadays some group theory is part of the stock-in-trade of almost every mathematician, but the same cannot be said of quasigroup theory. The paper by C.C. LINDNER, written in a very readable style, is recommended to anyone wishing to learn something about the subject.

Reference to the indices will show that although the papers are all on different topics some ideas and some names appear in more than one of them. In some cases the definition of a term used but left undefined by one author may be found in one of the other papers.

In editing this volume I have received assistance from the referees of the papers and from many other colleagues, to all of whom I express my grateful thanks. In particular my eagle-eyed colleague Gareth Jones read some of the papers and helped me to minimise the number of misprints and other errors in the volume now before you. The secretarial staff of the Faculty of Mathematical Studies, University of Southampton, and especially Margaret Youngs with her speedy and accurate typing, also provided valuable help. The London Mathematical Society agreed to the inclusion of this volume in their Lecture Note Series and also provided some financial assistance for the Conference. A book such as this has to be produced to a tight schedule and so I am very happy to thank the eight authors for their cooperation in producing their papers within the necessary time limits, and to thank Simon Capelin who has handled the editorial arrangements at Cambridge University Press.

Finally my thanks to Norman Pearce for his smooth driving of the coach on which most of this preface was composed during a weekend visit to study the industrial archaeology of Kent and East Sussex.

A special issue of 'Ars Combinatoria' will contain contributed papers of the Conference.

Hastings,
26 March 1983 E.K.L.

The paper by V.T. SÓS, Irregularities of partitions, was received at a very late stage and it has not been possible for it to be edited as thoroughly as the other papers nor for it to be indexed. The paper discusses the relationship between some classical results on uniformly distributed sequences and Ramsey-type theory. Some addenda and errata to the paper have been added on page 246.

Southampton,
13 April 1983 E.K.L.

GRAPHS AND INTERCONNECTION NETWORKS: DIAMETER AND VULNERABILITY

J.C. Bermond, J. Bond, M. Paoli & C. Peyrat
Université de Paris-Sud, Laboratoire de Recherche en Informatique, E.R.A. 452 du C.N.R.S., 91405 Orsay, France.

Contents

1 INTRODUCTION

It is well known that telecommunication networks or interconnection networks can be modelled by graphs. Recent advances in technology, especially the advent of very large scale integrated (VLSI) circuit technology have enabled very complex interconnection networks to be constructed. Thus it is of great interest to study the topologies of interconnection networks, and, in particular, their associated graphical properties. If there are point-to-point connections, the computer network is modelled by a graph in which the nodes or vertices correspond to the computer centres in the network and the edges correspond to the communication links. When the computers share a communication medium such as a bus, the network is modelled by a hypergraph, where the nodes correspond to the computer centres and the (hyper)edges to the buses. Note that there exists a second important class of networks, the "multistage networks", but we will not consider them. For a survey of interconnection networks, we refer the reader to Feng (1981).

In the design of these networks, several parameters are very important, for example message delay, message traffic density, reliability or fault tolerance, existence of efficient algorithms for routing messages, cost of the networks, ...

One important measure of the power of an interconnection network is the length of the longest path that the messages must travel from one node to another in the network, i.e. the distance between the nodes. It is advantageous to make these distances as small as possible, since this will reduce the message delay and also the message traffic density in the links. Worst case distance corresponds to the diameter of the associated graph or hypergraph. Similarly other network characteristics correspond to parameters of graphs or hypergraphs, e.g. mean distance, symmetry, connectivity, ...

A direct approach to network construction is to consider the graph model of possible links, with lengths and costs. The problem of designing a minimal or even near-minimal diameter subgraph with limited total cost was proved to be NP-complete by Plesník (1981). Also, Yannakakis (1978) proved that determining a connected subgraph with bounded maximal degree (the case of most network applications) of a given graph is NP-complete.

The object of this paper is to survey the results concerning diameter and connectivity in graphs and hypergraphs, in particular those of some importance for interconnection networks. For other results on the diameter we refer the reader to Bermond & Bollobás (1981), and on the connectivity to Mader (1979). Furthermore, we will only consider the deterministic aspect of these problems, though it is worth noting that the probabilistic aspect is considered in many papers and is of great importance for the reliability properties of networks.

Definitions and notation

We use standard terminology (see for example Berge (1973)). We give below some important definitions and notation which might be different from the usual terms.

The graph $G = (X,E)$ has vertex-set X and edge-set E. We denote by $n = |X|$ the number of vertices and by $m = |E|$ the number of edges. The degree $d(x)$ of a vertex x is the number of edges incident with x. We denote by δ the minimum degree and by Δ the maximum degree of the graph. A path with endpoints x and y is called an x-y path. The distance between two different vertices x and y is the length of a shortest x-y path. The diameter is the maximum distance over all the pairs of vertices. It will be denoted by D.

2 (Δ,D)-GRAPHS

In interconnection networks a practical problem is to interconnect the maximum number of nodes, while minimizing the diameter and knowing that, because of technical constraints, a node must not be incident to more than a fixed number of links. When a graph is associated with the network, this problem, first posed by Elspas (1964) and known as the (Δ,D)-graph problem, can be stated in graph-theoretic terms. How many vertices can a (Δ,D)-graph have, where a (Δ,D)-graph is a graph with maximum degree Δ and diameter at most D ? Let n(Δ,D) be the maximum number of vertices of a (Δ,D)-graph.

The Moore bound and Moore graphs

A theoretical bound on n(Δ,D) was given by Moore:

$$n(2,D) \leq 2D+1$$

$$n(\Delta,D) \leq \frac{\Delta(\Delta-1)^D-2}{\Delta-2} \quad \text{if} \quad \Delta \geq 3 .$$

Let the theoretical Moore bound be denoted $n_o(\Delta,D)$. The (Δ,D)-graphs with $n_o(\Delta,D)$ vertices are called Moore graphs. This name was given by Hoffman & Singleton (1960), because E.F. Moore raised the question with Hoffman and thought that eigenvalue techniques could be used to solve the problem (personal communication of Hoffman).

We warn the reader of the fact that the term Moore graphs is also used for what are known as cages. A cage is a Δ-regular of girth g having the minimum possible number of vertices. It can be shown that if g is odd, g = 2D+1 , the minimum number of vertices is also $n_o(\Delta,D)$ and the cages are of diameter D . Therefore in this case, cages are exactly Moore graphs. If g is even, g = 2D , the minimum number $n_1(\Delta,g)$ of vertices is

$$n_1(\Delta,g) = 2D \text{ if } \Delta = 2 ,$$

$$n_1(\Delta,g) = \frac{2(\Delta-1)^D-2}{\Delta-2} \quad \text{if} \quad \Delta \geq 3 .$$

We propose to call these graphs bipartite Moore graphs because they achieve the maximum number of vertices of a bipartite (Δ,D)-graph. For a survey on cages see Wong (1982).

It has been proved by various authors (see Biggs' book (1974)) that the Moore graphs exist only for Δ = 2 (2D+1-cycles), or D = 1 (Δ+1-cliques), or D = 2 and Δ = 3 (the Petersen graph), or D = 2 and $\Delta = 7$ (the Hoffman-Singleton graph) and possibly for $D = 2$ and $\Delta = 57$. Such a (57,2)-Moore graph has not been exhibited, however Aschbacher (1971) has shown that it cannot be distance transitive. Singleton (1966) proved that bipartite Moore graphs cannot exist unless $\Delta = 2$ (2Δ-cycles) or $D = 2$ (the complete bipartite graph $K_{\Delta,\Delta}$) or $D = 3$, 4 or 6. In the cases $D = 3$, 4 or 6, it has been shown, in particular by Benson (1966), that such cages (bipartite Moore graphs) exist if $\Delta = q+1$, q being a prime power. Therefore we know infinite families of bipartite Moore graphs.

Instead of undirected graphs, we can consider directed ones. If D is a digraph with indegree and outdegree at most Δ and diameter exactly D, the number of vertices is no greater than the <u>directed Moore bound</u>: $n_{Dir}(\Delta,D) = 1+\Delta+ \ldots + \Delta^D$. Bridges & Toueg (1980) have proved the impossibility of Moore digraphs (i.e. digraphs achieving the Moore bound) for $D > 1$ and $\Delta > 1$. This was also proved by Plesník & Znám (1974).

<u>Large (Δ,D)-graphs</u>

When Moore graphs do not exist, the determination of the exact value of $n(\Delta,D)$ appears to be a very difficult problem. So an interesting problem, especially for practical applications, is to find large lower bounds for $n(\Delta,D)$. For example, Erdös, Fajtlowicz & Hoffman (1980) ask the following question: given a non-negative number σ, is there a graph G with diameter 2, degree Δ and $\Delta^2+1-\sigma$ vertices? They show that it is false for $\sigma = 1$ except if $G = C_4$, and therefore $n(\Delta,2) \leq n_o(\Delta,2)-2$ if $\Delta \geq 4$ and $\Delta \neq 7$ or 57. In fact, this problem seems very difficult. A simpler one is to ask whether, for a fixed diameter D, there exist infinite families of (Δ,D)-graphs with about $C(D)\Delta^D$ vertices. That would mean $\liminf_{\Delta\to\infty} n(\Delta,D)\Delta^{-D} \geq C(D)$. We know by the Moore bound that $C(D) \leq 1$, and a question is: can we have $C(D) = 1$? Partial results on this problem will be given later on.

In order to give a lower bound for $n(\Delta,D)$, we have to show the existence of (Δ,D)-graphs with many vertices. For this purpose, there are basically two methods: use random graphs or design such graphs. For applications in interconnection networks, the second method is certainly more interesting than the first one since it really gives good graphs, whereas the first one, though perhaps more powerful, guarantees only their existence. However, we will note later on a few results on diameter in random graphs; for more details we refer the reader to

Klee & Larman (1981), Bollobás (1981) and Bollobas & de la Vega (1983).

Construction of large (Δ,D)-graphs

We will now survey the different types of method which have been used for the construction of large (Δ,D)-graphs.

Geometric methods: for instance, the existence of projective planes of order q, when q is a prime power, permits us to construct graphs with degree $\Delta = q+1$, diameter $D = 2$ and q^2+q+1 vertices. For these methods, see Delorme (1983a).

Direct methods: a simple example is given by the generalized de Bruijn graphs (de Bruijn (1946)). Let $q > 0$, $m \geq 2$ and $Q = \{0,1,\dots,q-1\}$. The de Bruijn graph $G = (Q^m,E)$ is the graph with vertex-set Q^m and edge-set E where $(\alpha,\beta) \in E$, if and only if $\alpha = a\gamma$ and $\beta = \gamma b$ for some $a,b \in Q$ and $\gamma \in Q^{m-1}$. It has maximum degree $\Delta = 2q$, diameter $D = m$ and q^m vertices. Since $q^m = \left(\frac{\Delta}{2}\right)^D$, we have: $\lim\limits_{\Delta\to\infty} N(\Delta,D)\Delta^{-D} \geq 2^{-D}$. These graphs have been generalized by Lam & van Lint (1978), Memmi & Raillard (1982), Delorme & Farhi (1983) and Quisquater (1983a).

Recently, we received a copy of Fiol, Alegre & Yebra (1983). They construct regular digraphs of out- and indegree Δ, diameter D, having $\Delta^D+\Delta^{D-1}$ vertices. In the case $D = 2$, these graphs have the maximum possible number of vertices $(\Delta^2+\Delta)$ because Moore digraphs (with $\Delta^2+\Delta+1$ vertices) do not exist. If we consider the associated undirected graphs, we have, for Δ even, $n(\Delta,D) \geq \left(\frac{\Delta}{2}\right)^D + \left(\frac{\Delta}{2}\right)^{D-1}$. The authors also show that their digraphs can be viewed as iterated line digraphs of complete symmetric digraphs. Still more recently, Y. Raillard informed us that Kautz (1969) used these same graphs, which he referred to as state diagrams of shift registers, for the same purpose.

Addition of vertices: in certain graphs, vertices can be added without changing the degree or the diameter of the initial graph. See for example Delorme (1983a, 1983b). Recently, one of us (Bond) has shown that vertices can be added to the de Bruijn graphs. For there is a loop in these graphs for each vertex $(a,a,\dots,a)$ and a double-edge between two vertices of the form $(a,b,a,b,\dots)$ and $(b,a,b,a,\dots)$. Eliminating these loops and double-edges, there remain q points of degree $2q-2$ and $q(q-1)$ of degree $2q-1$. Then we add q points x_i, $i = 0,1,\dots q-1$, with q edges between x_i and vertices $(\alpha,i,\alpha,i,\dots)$ for any $\alpha \in Q$, so that x_i has degree q. We add a vertex x_q (of degree q) connected to $(\alpha,\alpha,\dots,\alpha)$ for any $\alpha \in Q$. So we can add q other vertices y_i $(i \in Q)$

such that they form a complete bipartite graph $K_{q+1,q}$ with the $(q+1)$ vertices x_i, $i = 0,1,\ldots,q$. Now, we have a graph with q^m+2q+1 vertices, all of degree $2q$ except y_i. The resulting completed de Bruijn graph still has diameter m.

Products of graphs: in general, the classic products do not give good results, but Bermond, Delorme & Farhi (1982, 1983) have defined a new product which produces a lot of interesting (Δ,D)-graphs for small values of D and Δ.

Delorme (1983b) considers a product which, from two bipartite graphs, gives a bipartite graph. It consists of taking one of the connected components of the Kronecker product of two graphs; in what follows we will call it also a Kronecker product.

Let us consider two bipartite graphs $(X \cup X',E)$ and $(Y \cup Y',F)$ with diameters D and D' respectively, and degrees Δ_X, $\Delta_{X'}$, Δ_Y and $\Delta_{Y'}$ for the vertices of X,X',Y and Y' respectively. The Kronecker product of these two graphs has vertex set $(X \times Y) \cup (X' \times Y')$, and (x,y) is adjacent to (x',y') if and only if $(x,x') \in E$ and $(y,y') \in F$. The new graph is bipartite with maximum degree $\max(\Delta_X\Delta_Y,\Delta_{X'}\Delta_{Y'})$ and diameter $\max(D,D')$. Thus, if we take the Kronecker product of a bipartite graph $G = (X \cup X',E)$ and its opposite $(X' \cup X,E)$, the resulting graph has degree $\Delta_X\Delta_{X'}$ and diameter $D(G)$.

Delorme uses for $(X \cup X', E)$ the bipartite Moore graph with diameter 4 and degree $(q+1)$, where q is a prime power, and with $|X'| = |X| = (q+1)(q^2+1)$. He obtains thus a bipartite graph $P = (A \cup B,\mathcal{E})$ of diameter 4, degree $(q+1)^2$, and $2(q+1)^2(q^2+1)^2$ vertices. This graph P has a polarity, i.e. an automorphism f such that $f(A) = B$ and $f(B) = A$ and $f^2 = id$; here $f((x,x')) = (x',x)$. The quotient of this graph P by the polarity is the graph with vertex set A where two distinct vertices a and a' are adjacent if and only if a and $f(a')$ are adjacent in P. The quotient Q has degree $(q+1)^2$, $(q+1)^2(q^2+1)^2$ vertices and diameter D equal to 3. Thus, we obtain a family of graphs of degree Δ, diameter 3 and with about Δ^3 vertices. Therefore: $\liminf_{\Delta\to\infty} n(\Delta,D)\Delta^{-3} = 1$. (We recall that because of the Moore bound $\liminf_{\Delta\to\infty} N(\Delta,D)\Delta^{-D} \leqslant 1$). In the same way, using the bipartite Moore graphs of degree $q+1$ and diameter 6, Delorme obtained by the same methods a family of graphs with degree Δ, diameter 5 and about Δ^5 vertices. For $D = 1,2,\ldots,10$ he gave improved lower bounds u_D for the asymptotic values $\liminf_{\Delta\to\infty} N(\Delta,D)\Delta^{-D}$.

D	1	2	3	4	5	6	7	8	9	10
u_D	1	1	1	$3^3.2^{-7}$	1	$2.5^6.6^{-6}$	$6^6.7^{-7}$	3.2^{-8}	14.3^{-8}	5.2^{-10}

Compounding methods: they have been developed by Bermond, Delorme & Quisquater (1982b), and also by Uhr (1983). In general, they consist of replacing the vertices in a large graph by copies of the same small graph or by different graphs. Some of these constructions were also considered by Wegner (1976); he proved, for example, that $n(\Delta,D) \geq (\Delta+1)n(\Delta-1,D-2)$.

Jerrum & Skyum (1983) have compounded de Bruijn graphs with a graph G', which have good properties of average path length, to obtain for Δ fixed, graphs with a diameter equal to $\lambda_\Delta \log_2 n+O(1)$. Their values of λ_Δ, which depend on Δ only, improved those of the de Bruijn graphs and are now the best obtained by constructive means. For example, they obtained cubic graphs with diameter $1.472169 \log_2 n+O(1)$; a previous construction giving a diameter of $1.5 \log_2 n+O(1)$ was given by Leland & Solomon (1982). However, the results on random graphs of Bollobas and de la Vega (1983) give better values. One of their main results is that, for $\Delta \geq 3$ and $\varepsilon > 0$ fixed, if $D = D(n)$ is the least integer satisfying $(\Delta-1)^{D-1} \geq (2+\varepsilon)\Delta n \log n$, then almost every Δ-regular graph of order n has diameter at most D. As a corollary of this result, for a fixed degree Δ, the minimum diameter of a graph of order n is bounded above by $\frac{\log n}{\log(\Delta-1)} + O(\log \log n)$. This is the best possible result because of the Moore bound: $D \geq \frac{\log n}{\log(\Delta-1)}$. It would be interesting to construct families of graphs which attain this bound.

Computer searches: these have been done by Quisquater (1983a) using Y-graphs and recently by Doty (1982) using chordal ring. Other algorithmic methods which need the help of a computer have been developed before by Toueg & Steiglitz (1979), Imase & Itoh (1981) and Arden & Lee (1982). Different tables of the largest known (Δ,D)-graphs, for $\Delta \leq 15$, $D \leq 10$, have been given by Storwick (1970), Leland, Finkel, Qiao, Solomon & Uhr (1981), Memmi & Raillard (1982) and Bermond, Delorme & Quisquater (1982a).

Minimum number of edges of a (Δ,D)-graph

Another problem close to the (Δ,D)-graph problem was proposed by Erdös & Rényi (1962). Suppose that there is a (Δ,D)-graph of order n. Let $e_D(n,\Delta)$ be the minimum number of edges of such a graph. An interconnection network whose associated graph achieves this bound is interesting because it has a minimum cost. Bollobás (1978, ch IV) presented, in the case $D \leq 3$, results of Erdös & Rényi (1962), Erdös, Rényi & Sós (1966) and Bollobás (1971). More recently, the case $D = 2$ has been settled almost completely by Pach & Surányi (1981). They defined a function $g(c)$ on $[0,1]$ and proved that $\lim_{n\to\infty} e_2(n,\lfloor cn \rfloor)/n = g(c)$ for every c, $0 < c < 1$, except for $c = c_1, c_2, \ldots$ where (c_k) is a sequence tending to 0.

Mean distance

We may also consider that the important parameter in an interconnection network is not the maximal time spent to travel from one node to another, but the average time. Therefore, we must study instead of the diameter of the associated graph, its average path length or mean distance. It has been defined by Doyle & Graver (1977):

$$\mu(G) = \frac{1}{\binom{n}{2}} \sum_{v,w \in X} d(v,w) .$$

Few things have been done on this parameter except for certain particular cases. The mean distance is easily calculated if the graph has great properties of symmetry, for example if it is distance degree regular, which means that each vertex has exactly D_j vertices at distance j from it. As this sort of graph is far from our topic, for other properties of them, we refer the reader to a survey of Bloom, Kennedy & Quintas (1981). Buckley & Superville (1981) have determined the mean distance for various classes of graphs, and Buckley (1981) for their line-graphs.

Cerf, Cowan, Mullin & Stanton (1975) introduced a problem similar to the (Δ,D)-graph problem: given the number n of vertices and the degree Δ of a graph, what is the lower bound of the mean distance? The graphs which achieve this bound are called generalized Moore graphs. The same authors (1974a, 1974b, 1976) described these graphs or established their non-existence up to $n = 34$, for $\Delta = 3$.

Other problems of distance have been investigated in the literature, for example problems of radius, eccentricity or of "locations in networks" (see the survey of Tansel, Francis & Lowe (1983) which contains many references).

3 CONNECTIVITY

In the design and analysis of interconnection networks, one of the fundamental considerations is the reliability, in particular that the centres can communicate in case of link or node failure. Reliability has been defined in a number of different ways, either with deterministic or probabilistic criteria. Maybe the probabilistic criteria fit better with the real world, indeed vertices or links can fail randomly: one of the main problems is to compute the probability that there is an operating path between two nodes knowing the probability of vertex or link failure. We will not discuss this non-combinatorial aspect here; we refer the reader to the surveys of Frank & Frisch (1970, 1971), Wilkov (1972) and Hwang (1978) in the case of multistage networks. Note that Ball (1980) has recently shown that virtually all network reliability analysis problems of practical interest are NP-hard.

In the case of deterministic networks, the criteria indicate how difficult it is to disrupt the operation of the whole network (or some part of it). Therefore one aim is to maximize the number of nodes or links that must fail in order to disrupt the operation of the network, while taking into account that there are cost constraints. The main criteria are the classical connectivity or edge-connectivity and the local connectivity.

Classical results

Recall that an x-y *separating set* S is a non-empty subset of vertices of G whose deletion destroys all the paths between x and y. The *local connectivity* between x and y is defined as the minimum cardinality of an x-y separating set. The *connectivity* of a non-complete graph is the minimum value of the local connectivity over all pairs of vertices. The connectivity will de denoted by $\kappa(G)$. By convention we set $\kappa(K_n) = n-1$. Analogous definitions can be given for local edge-connectivity and edge-connectivity; the latter will be denoted by $\lambda(G)$.

For an excellent survey of connectivity see Mader (1979). We will mention here only a very small number of results, which are of interest for our purpose. Recall first that by Menger's theorem the local connectivity between x and y is equal to the maximum number of vertex-disjoint paths between x and y. A similar result holds with edges instead of vertices. Using this or flow techniques there exist good (i.e. polynomial) algorithms to determine the local or the global connectivity (or edge-connectivity) of a graph (see for example Even's book (1979) or Chvátal's book (1983)). An algorithm in $O(n^3)$ for finding maximum flows is given in Malhotra, Kumar & Maheshwari (1978).

Symmetric graphs

Another well-known result says that, in any graph G, $\kappa \leq \lambda \leq \delta$. When the graph has some symmetries the following results are interesting; in particular they ensure that the graph has the best connectivity possible without any computation. A graph is said to be vertex-transitive if for every pair of vertices x and y there exists an automorphism f of the graph such that $f(x) = y$. Edge-transitive graphs are defined analogously.

Watkins (1970) and Mader (1970) have shown that if G is a connected edge-transitive graph then $\kappa = \lambda = \delta$. In the case of connected vertex-transitive graphs they have proved that $\kappa \geq 2\lfloor\frac{\delta}{3}\rfloor + 2$. Furthermore Mader (1971) proved that in any connected vertex-transitive graph $\lambda = \delta$. Mader (1970) also showed that if G is a connected vertex-transitive graph without a complete graph on 4 vertices then $\kappa = \delta$.

This result of Mader (1970) can be used to show that each of the graphs defined by Memmi & Raillard (1982) (see §2) has its connectivity equal to its degree. This result has been proved directly by exhibiting the paths by Amar (1983).

Maximum connectivity

Note that as $\kappa \leq \lambda \leq \delta$ the maximum connectivity or edge-connectivity of a graph with n vertices and m edges is $\lfloor\frac{2m}{n}\rfloor$. Harary (1962) exhibited constructions of graphs having this maximum connectivity. In fact, that follows from the construction of graphs on n vertices, of connectivity κ, having the minimum number of edges (and corresponds to the problem of designing a minimum cost network, in which all the edges have the same cost with a uniform reliability). If the connectivity k is even, the best known of these graphs is $C_n^{\frac{1}{2}k}$ the $\frac{1}{2}k$ power of a cycle C_n (that is the graph in which two vertices i and j are joined if $|j-i| \leq \frac{1}{2}k\ [\mathrm{mod}\ n]$). If the connectivity k is odd and the number of vertices is even, we take $C_n^{\frac{1}{2}(k-1)}$ plus the diameters (edges between i and $i + \frac{1}{2}n[\mathrm{mod}\ n]$) . The same construction works in the case k odd, n odd, but one vertex is of degree $k + 1$.

Recently it has been proved by Paoli, Wong & Wong (1983) that these graphs (or some variant of them) are also optimal for designing a minimum cost network with the property that even after k links (or nodes) have failed it still can function as a basic ring network. In graph theoretical terms this means that the above graphs are k-edge (vertex) hamiltonian.

Optimization problems

For constructions and properties of the graphs with maximum connectivity see either Wilkov's survey (1972) or Boesch's survey (1983). Note that the construction of minimum graphs on n vertices with connectivity κ is a particular case of this generalized synthesis problem: given a weighted graph on n vertices in which every edge e has a cost $c(e)$ and given an integer $n \times n$ matrix of required vertex (edge) connectivities U_{ij} , find a partial graph of G of minimum total cost such that the local connectivity between vertices x_i and x_j is at least U_{ij} $(1 \leqslant i < j \leqslant n)$.

The case considered above deals with a uniform cost and $U_{ij} = \kappa$. There is a wide literature on this general problem, but we will not deal with it as it has been surveyed very recently by Christofides and Whitlock (1981).

Connectivity and diameter

Unfortunately most of these graphs, in particular those constructed as powers of cycles have a great diameter. For example $C_n^{\frac{1}{2}k}$ has diameter $\lceil \frac{n}{k} \rceil$. Therefore it would be of interest to construct maximally connected graphs with the smallest diameter possible. For example, an important problem is to build regular graphs with connectivity equal to the degree Δ and minimizing the diameter D , which, because of the Moore bound, is at least $\frac{\log n}{\log(\Delta-1)}$. Since this lower bound is very sporadically achieved, only by the Moore graphs, Wilkov (1972) asked for an algorithm doing this construction for every n ; but this seems to be a very difficult problem. However, Schumacher (1983) has given a polynomial algorithm (complexity $O(n^2)$) to construct Δ-regular, Δ-connected graphs on n vertices with diameter at most twice the Moore bound.

Generalizations

Finally let us note that here we have considered essentially as the criterion of reliability the classical connectivity or edge-connectivity.

One objection to using only these parameters is that they fail to differentiate between the different types of disconnected graphs which result from removing κ vertices or λ edges. It is not the same in practice to isolate a single vertex or to divide the graph into two equal components. Several generalized measures of connectedness have been proposed in the literature. A recent complete survey of these parameters has been done by Boesch (1983). We will see in the next paragraphs some measures involving connectivity and diameter.

4 DISJOINT PATHS

Mengerian-type problems

In the reliability analysis of a communication network it is often required to find the maximum number of vertex-disjoint paths between two given vertices of the network with the requirement that the length of each path does not exceed a given value. With this additional constraint the problem becomes more complicated than the k-connectivity.

Let $I_k(x,y)$ be the maximum number of vertex-disjoint paths between x and y of length not greater than k. Itai, Perl & Shiloach (1982) have shown that the determination of $I_k(x,y)$ is an NP-complete problem if $k \geq 5$. This fact is not surprising because, for $k \geq 5$, there does not exist a Mengerian-type theorem.

Let $S_k(x,y)$ be the minimum number of vertices whose deletion destroys all x-y paths of length not greater than k. We always have $I_k(x,y) \leq S_k(x,y)$, but in general equality does not hold.

For $k \geq 5$ Lovász, Neumann-Lara & Plummer (1978) have exhibited graphs with $I_k(x,y) < S_k(x,y)$. In fact they have shown that

$$\sqrt{\frac{k}{2}} \leq \operatorname*{Sup}_G \frac{S_k(x,y)}{I_k(x,y)} \leq \left\lfloor\frac{k}{2}\right\rfloor .$$

The graph of figure 1 has $I_5(x,y) = 1$ and $S_5(x,y) = 2$.

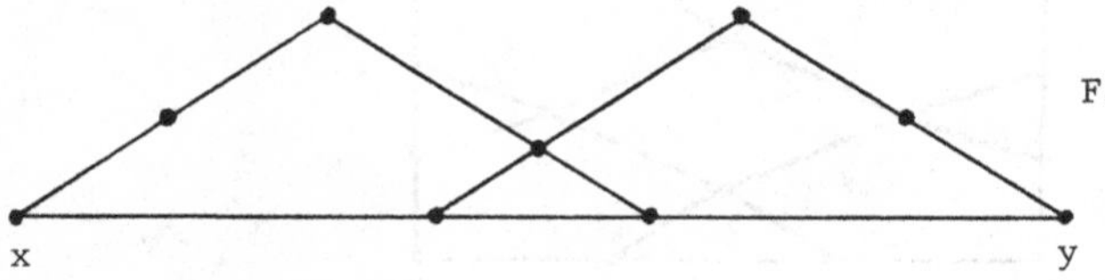

Fig. 1

More recently Boyles & Exoo (1982) found that $\mathrm{Sup}\frac{S_k(x,y)}{I_k(x,y)} \geq \lfloor\frac{k+3}{4}\rfloor$, thereby disproving a conjecture of Lovász and Neumann-Lara mentioned in Bollobás' book (1978) (ex n° 17, p.48) and in Bondy & Murty (1976) (conjecture n° 17). Chung (1983) improved the lower bound to $\frac{1}{3}(k+1)$ and asks whether $\mathrm{Sup}\,\frac{S_k(x,y)}{I_k(x,y)} = \frac{1}{3}k + O(1)$.

However if $k \leq 4$ we have $I_k(x,y) = S_k(x,y)$ and the determination of $I_k(x,y)$ is a polynomially solvable problem.

We remark that if $k \geq n-1$ then $I_k(x,y) = S_k(x,y)$ is nothing else but Menger's theorem. In the particular case where k is the distance between the two vertices we also have:

$$I_k(x,y) = S_k(x,y) \qquad k = d(x,y).$$

This result is due independently to Entringer, Jackson & Slater (1977) and Lovász, Neumann-Lara & Plummer (1978). Itai, Perl & Shiloach (1982) have shown that the determination of the maximal number of vertex-disjoint paths of shortest length is a polynomially solvable problem. By applying flow techniques the algorithm requires $O(|V|^{\frac{1}{2}}|E|)$ time. Itai, Perl & Shiloach (1982) and Hartman & Rubin (1978) have found analogous results for edge-disjoint paths.

The above results concern local properties of the graph. It would be interesting to measure how the deletion of edges or vertices increases the diameter of the graph. A parameter for this purpose has been proposed by Boesch, Harary & Kabell (1981) who defined the <u>persistence</u> $\rho_0(G)$ as the minimum number of vertices which must be removed in order to increase the diameter or yield a trivial graph.

Unfortunately, contrary to the claim in their article, the equality $\rho_0(G) = \mathrm{Inf}\, S_{D(G)}(x,y)$ is not true. Indeed, Bondy & Hell (1983) have found an infinite family of counterexamples such that $\rho_0(G)$ is strictly greater than $\mathrm{Inf}\, S_{D(G)}(x,y)$ (see also Exoo (1982)). Figure 2 is a counterexample with $D(G) = 5$, $\rho_0(G) = 2$ and $S_5(x,y) = 1$.

Fig. 2

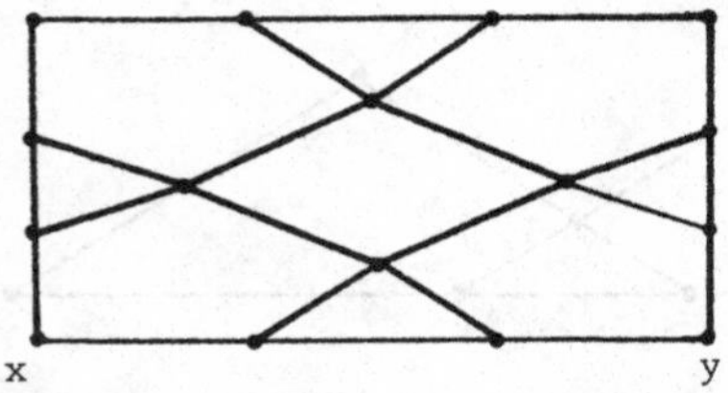

An interesting problem due to Elspas is to determine the maximum number of vertices of graphs of maximal degree Δ such that each pair of vertices are joined by μ edge-disjoint paths of length at most k (where $I_k(x,y) = \mu$) . Some particular cases of this problem have been considered by Quaife (1969).

Geodetic graphs

Instead of searching for graphs with a great number of disjoint paths between vertices, some authors have studied graphs in which any two vertices x and y of V are joined by exactly one shortest path. Such graphs are called geodetic graphs. In some sense these graphs are very vulnerable and therefore less interesting for applications. A survey of the main results and open problems concerning geodetic graphs can be found in Bosák (1978). Geodetic graphs homeomorphic to a complete graph were characterized by Stemple (1979).

A graph G is said to be strongly geodetic if each pair of vertices is connected by at most one path of length less than or equal to the diameter of G. Bosák, Kotzig & Znám (1968) proved that a strongly geodetic graph is either a forest (union of trees) or a Moore graph.

5 DIAMETER VULNERABILITY AND EXTREMAL PROBLEMS

Deletion and augmentation problems

Plesník (1975a) has shown that the deletion of an edge in a graph of diameter D gives either a disconnected graph or a graph of diameter at most $2D$. This bound was greatly generalized by Chung and Garey (1983). They have shown that if s edges are removed from a graph of diameter D , the resulting graph, if it is connected, has diameter at most $(s+1)D + O(s)$. To prove this result they showed that the least possible diameter D' of a graph obtained by adding s edges to the path on n vertices satisfies

$$\frac{n}{s+1} - 1 \leq D' < \frac{n}{s+1} + 3 .$$

They also considered the same problem for cycles.

An interesting problem would be to consider the same augmentation problem with the maximum degree of the resulting graph fixed. For example, the following natural question due to Farley & Hedetniemi (1978) is of interest: given a cycle of length $2n$, what is the least possible diameter of a cubic graph obtained by adding a matching of n edges?

These augmentation problems are also of interest when considering adding facilities to a network in order to upgrade its performance. In this spirit, as asked by Chung and Garey, it would be interesting to determine the computational complexity of the following problem: given s, D and G, determine whether there exists a supergraph of G obtained by adding s edges (vertices) that has diameter no more than D and, if so, find such a supergraph. And if this problem is NP complete, do good heuristics exist?

Chung and Garey also studied the vertex-deletion case; however in this case the corresponding maximum diameter is unbounded in terms of s and $D(G)$; nevertheless they proved that the removal of s vertices in a k-connected graph $(k > s)$ gives a graph of diameter at most $\frac{n-s-2}{k-s} + 1$. Similarly it would be interesting to know the complexity of the following problem: given s, D and G determine whether there exists a subgraph of G obtained by deleting s edges (vertices) that has diameter no more than D and, if so, find such a subgraph.

(Δ,D,D',s)-graphs

A related problem is to find $N(\Delta,D,D',s)$, the maximum order of a (Δ,D,D',s)-graph, i.e. a (Δ,D)-graph such that the maximum diameter of the graph obtained from G by deleting s vertices is D'. Some values of $N(\Delta,D,D',s)$ were obtained by Bond, Moulinoux & Peyrat (1983). They obtained bounds for $N(\Delta,2,D',1)$ and, for "large" graphs of diameter 2, proved that $D' = 3$. They also showed that $N(\Delta,D,D,1) \geq (\Delta/2)^D + (\Delta/2)^{D-1}$.

In order to find graphs of diameter D such that after the deletion of s vertices the diameter remains small, one can consider graphs of diameter D and girth $2D$, such that every two vertices at distance D are connected by $s+1$ vertex-disjoint paths. The existence of some of these graphs is studied in Gewirtz (1969). For $s=1$ see also Roos & van Zanten (1983c).

An extremal problem

Murty & Vijayan (1964) proposed the problem of minimizing the number $f_v(n,D,D',s)$ (resp. $f_e(n,D,D',s)$) of edges in a graph of order n and diameter D and having the property that if any s vertices (resp. edges) are deleted, the remaining graph has diameter at most D'. These functions have been studied by many authors, but their determination seems a difficult problem in general: essentially the cases solved are $s = 1$ or $D \leq 4$; see the survey of Caccetta (1979). Recently Usami (private communication of Enomoto) has shown that $f_v(n,D,6,1) = \max\left\{n, \left\lceil\frac{4n-8}{3}\right\rceil\right\}$ for $4 \leq D \leq 6$. The asymptotic behaviour of $f_v(n,D,D',S)$ for large D' is known; it has been proved that if $D \geq 2$ and $D' \geq 2D - 2$ then $\lim \frac{f_v(n,D,D',1)}{n} = \frac{D}{D-1}$ (see Bollobás' book (1978 ch. IV 2)). Recently, Enomoto & Usami (1983) have determined the exact value by showing that $f_v(n,D,D',1) = \left\lceil\frac{Dn-2D-1}{D-1}\right\rceil$ if $n > D' \geq 2D - 1$. This settles a conjecture of Caccetta (1979), and as an immediate corollary, determines the minimum number of edges of a 2-connected graph of diameter D (in the case $D' = n-1$). The same result seems to have been obtained by Suvikov (1978) according to Mathematical Reviews (80C 05095). Finally, this last result settles a conjecture of Glivjak (1976) by proving that the minimum number of edges of an equi-essentric graph of diameter D is $\left\lceil\frac{Dn-2D-1}{D-1}\right\rceil$ (an <u>equi-eccentric graph</u> is a graph in which $\max_{y\in G} d(x,y) = D$, for every x), for such a graph is 2-connected.

Enomoto (private communication, November 1982) has also shown that the minimum number of edges of a graph of order n, minimum degree 2 and diameter D is also equal to $\left\lceil\frac{Dn-2D-1}{D-1}\right\rceil$ if D is odd, $n > 2D$, but to $\min\left\{\left\lceil\frac{Dn-2D-1}{D-1}\right\rceil, \left\lceil\frac{(D+1)(n-1)}{D}\right\rceil\right\}$ if D is even and $n > 2D$. As a corollary, he determined $f_e(n,D,D',1)$ for large D'.

Note that the case $D' = n-1$ corresponds to the case where G is s-connected or s-edge-connected. In that sense the result of Chung and Garey mentioned at the beginning of this paragraph shows that, if $D' \geq (s+1)D+s$, then $f_e(n,D,D',s) = f_e(n,D,n-1,s)$. The case where D' is close to D, and in particular $D' = D$, is very far from being settled. Bollobás (1978, ch. IV 3) has conjectured that

$$\lim_{n\to\infty} \frac{f_e(n,D,D',1)}{n} = \frac{\lfloor\frac{1}{2}D\rfloor + 1}{\lfloor\frac{1}{2}D\rfloor}$$

. Bollobás & Eldridge (1976) have shown that

if $n \geq 2s$, then $f_v(n,2,2,s) = (s+1)(n-s-1)$. Entringer & Jackson (1982) have determined $f_v(n,2,2,s)$ when $n \leq 2s - 1$. This result is obtained in fact by determining the minimum number of edges of a graph with n vertices and diameter 2 , which is s-geodetically connected (i.e. a connected graph such that the removal of at least s vertices is required to increase the distance between any pair of vertices).

Diameter-critical graphs

A completely different point of view consists of considering vertex- (resp. edge-) critical graphs, i.e. graphs such that the deletion of any vertex (resp. edge) increases the diameter.

The characterization of these graphs seems a very difficult problem. Glivjak (1975) has proved the impossibility of a characterisation of edge-critical graphs by finite extension or by forbidden subgraphs. However partial results have been obtained (see for example Plesník (1975b)) and, in particular, Plesník and Murty and Simon have conjectured that any edge-critical graph of diameter 2 on n vertices has at most $\lfloor \frac{1}{4}n^2 \rfloor$ edges. Some progress has been made by Caccetta and Häggkvist (1979) who have obtained the bound $\frac{1}{12}(1 + \sqrt{5})n^2 < 0.27\ n^2$. A general conjecture on the maximum number of edges of an edge-critical graph of diameter D has been proposed by Krishnamoorthy & Nandakumar (1981) who disproved an earlier conjecture of Caccetta & Häggkvist (1979).

Other extremal problems

Another extremal problem consists of finding the maximum diameter of a graph with given minimum degree and connectivity. This has been solved by Amar, Fournier & Germa (1983) who determine the minimum order $f(D,\delta,\kappa)$ of a graph with diameter D , minimum degree δ and connectivity κ :

$$f(1,\delta,\kappa) = \delta+1$$

$$f(2,\delta,\kappa) = \delta+2$$

$$f(D,\delta,\kappa) = \begin{cases} 2(\delta+1) + (D-3)\kappa & \text{if } D \geq 3 \text{ and } 1 \leq \kappa \leq \delta \leq 3\kappa - 1, \\ (t+1)(\delta+1) + \varepsilon\kappa & \text{if } D \geq 3 \text{ and } \delta \geq 3\kappa - 1 , \end{cases}$$

where $D = 3t + \varepsilon$, $\varepsilon \in \{0,1,2\}$.

Analogous results for hypergraphs have been found by Raynal (1983).

Myers, Klee & Quaife, furthermore, studied the cases where D is δ-regular, in particular $\kappa = 1$ or $\Delta = 3$ (see Myers (1981) for a survey on these results). For digraphs see Hirschberg & Wong (1979).

Cayley graphs

Let $S_N = \{\Pi_1, \Pi_2, \ldots, \Pi_{N!}\}$ be the symmetric group of permutations on N elements. Let $\Sigma_r = \{\sigma_1, \ldots, \sigma_r\}$ be a system of r generators of S_N. The associated Cayley graph has:

(1) as vertices the N! permutations in S_N,

(2) two vertices, indexed by the permutations Π_i and Π_j in S_N, are adjacent if and only if $\Pi_i \sigma_k = \Pi_j$, for some $\sigma_k \in \Sigma_r$. By an enumerative argument, the diameter D is greater than $\log_r(N!)$. Quisquater (1983b) has improved results of Golunkov (1973) by showing that for each N and r, there exists a system Σ_r such that the diameter $D = c_r \log_r(N!)$. This result has some applications for dynamic memories. Other properties of Cayley graphs have been studied by Tchuente (1982).

6 HYPERGRAPHS

In this section, we consider the case where sets of processors share a communication medium such as a bus. As we said in the introduction, these networks are represented by hypergraphs: the vertices represent the processors and the (hyper-)edges the buses. Although these networks are now less used than the classical ones with point-to-point connections, their importance is growing and they might be very important in the near future. Our interest in this subject came from discussions with persons working at the French National Telecommunications Centre (CNET), on a project called REBUS. Examples of such networks are the well-known d-dimensional hypercubes of width r. The vertices can be seen as d-tuples $(a_1, \ldots, a_d)$, whose coordinates can take the values $0, 1, 2, \ldots, r-1$; therefore the total number of vertices is r^d. An edge consists of all the vertices having $d - 1$ coordinates fixed. So a vertex belongs to exactly d edges and an edge contains exactly r vertices. Particular cases are the binary hypercube $(r = 2)$ which is a graph, and the "grid" or "mesh" $(d = 2)$.

The grid structure is essentially that used in the computer interconnection systems Illiac IV or MICRONET (see for example the article of Wittie & Van Tilborg (1980)). Other such networks have been proposed by Finkel & Solomon (1980); Wittie (1981): the "Dual bus hypercubes", and by Finkel & Solomon (1981): the "Lens structured networks".

As we have already said in the introduction, the important parameters in such networks are the message delay, the reliability, the existence of efficient routing algorithms, etc., which correspond, as in the case of graphs, to parameters of the associated hypergraph. A technical constraint is that a node must not belong to too many buses; i.e. a vertex of the hypergraph has bounded degree. Furthermore, the traffic load on each bus is limited. Since, for general networks without symmetry properties, the traffic per bus is not easily calculated, we will restrict our study to networks with some uniformity and, to a first approximation, replace the constraint of traffic load by the limitation of the number of processors connected to each bus. This means that the size r of the edges of the associated hypergraph is bounded.

A comparison of the different criteria on a few structures has been done by Wittie (1981); however, it is impossible to optimize all the criteria at the same time. From a combinatorial point of view, a study of the extremal problems is worthwhile. We have begun such a study and our aim in what follows is to indicate some further directions of research; indeed as we shall see, most of the area is wide open.

Notation and definitions

Our notation and definitions will be:

- $H = (X,E)$, for a hypergraph with vertex-set X, and (hyper)edge-set E,
- the degree of a vertex is the number of edges containing it,
- Δ for the maximum degree over all vertices of H,
- r for the maximal size of an edge,
- a path in a hypergraph connecting x and y will be a sequence $x = x_1, E_1, x_2, \ldots, x_i, E_i, x_{i+1}, \ldots, E_{p-1}, x_p = y$ with $\{x_i, x_{i+1}\} \subset E_i$. Its length will be $p-1$, the number of edges of the path. Distance and diameter are then defined and denoted as in the case of graphs.

The (Δ,D,r)-hypergraph problem

A (Δ,D,r)-hypergraph will be a hypergraph of diameter D, maximum degree Δ and maximum edge-size r. The problem identical to the (Δ,D)-graph problem is to determine the maximum number of vertices $n(\Delta,D,r)$ of a (Δ,D,r)-hypergraph. As for graphs, $n(\Delta,D,r)$ is bounded above by a "Moore bound":

$$n(\Delta,D,r) \leqslant 1 + \Delta(r-1) \sum_{i=0}^{D-1} (\Delta-1)^i (r-1)^i .$$

In the case $r = 2$. we obtain the classical Moore bound.

The first problem is to determine if this bound can be reached. This problem has been considered by different persons independently (Kuich & Sauer (1969), Bose & Dowling (1971), Fuglister (1977), Damerell & Georgiacodis (1981a, 1981b)). It appears also as a particular case of more general structures. The hypergraphs which attain the bound are often called Moore geometries.

First look at the case $D = 1$, that is when each pair of vertices belongs to at least one edge. The reader will see the similarity with design theory. Let us recall that a (v,k,λ)-BIBD (balanced incomplete block design) is a collection of subsets (called blocks) of a given set on v vertices such that every block contains exactly k vertices and every pair of vertices belongs to exactly λ blocks. In fact, the bound $n(\Delta,1,r) = 1 + \Delta(r-1)$ is attained if and only if there exists an $(n,r,1)$-BIBD. Therefore this case is extensively studied: for values of the parameters for which there exists an $(n,r,1)$ design, see for example Hanani (1975) or Doyen & Rosa (1980). We will come back to the case $D = 1$ later.

For $D > 1$, successive efforts of different persons using techniques similar to those used for graphs, have proved that the bound can be attained only if $D = 2$. The case $D = 3$ has been excluded by Fuglister (1977), the case $D > 4$ by Damerell & Georgiacodis (1981b) and the case $D = 4$ by Ott (1983). The case $D = 2$, $r = 2$ corresponds to the Moore graphs; for $D = 2$, $r > 2$, it has been proved by Bose & Dowling (1971) (see also Kantor (1977)) that, if $r \neq 5$, Moore hypergraphs can exist only for a finite number of values. For $r = 3$, there cannot exist Moore hypergraphs. For $r = 4$, there might exist one Moore hypergraph with $\Delta = 7$, 400 vertices and 700 edges. For $r = 5$, there might exist Moore hypergraphs of degree s^2 with $s = 0, 1$, or $4 \pmod 5$. The cases $D = 2$, $r = 3$ and $r = 4$ have also been done by Kuich & Sauer (1969) who studied hypercages, i.e. regular hypergraphs of degree Δ, with uniform size r, of girth g and having the minimum possible number of vertices. In the case g odd, $g = 2D+1$, it can be shown that these hypergraphs have diameter D. They studied the case $g = 5$ (note that there is a small computational mistake in the last corollary for the case $r = 4$). However, it is not known if such Moore hypergraphs exist for $r \geqslant 3$.

We may also note that these Moore hypergraphs appear as particular cases of more general structures. Very recently, for example, Buekenhout (1983) has defined (g,d_p,d_L)-gons as bipartite graphs with the two disjoint vertex-sets P and L respectively of points and lines, with girth 2g, point diameter d_p and line diameter d_L where:

$$d_p = \max_{p \in P} \max_{x \in P \cup L} (d(p,x)) ,$$

$$d_L = \max_{\ell \in L} \max_{x \in P \cup L} (d(\ell,x)) .$$

Delorme (1983c) has also defined a class of bipartite graphs, generalizing distance-regular graphs. Particular cases of these graphs are related to hypergraphs in the following sense: to each hypergraph $H = (X,E)$, we associate a bipartite graph with vertex set $X \cup E$, where a vertex x is joined to an edge e if x belongs to e in H. The point diameter of the resulting graph is 2D or 2D+1, and the line diameter between 2D-1 and 2D+2.

Another generalization of Moore hypergraphs is due to Roos & van Zanten (1983a,b). They consider Δ-regular, r-uniform hypergraphs such that any pair of vertices at distance strictly less than D are connected by a unique shortest path, and pairs of vertices at distance D by a constant number of paths of length D.

Case D=1

Let us come back to the case D = 1 : Moore hypergraphs exist if and only if there exist (n,r,1)-BIBDS, but in this case $\Delta \geqslant r$. The case $\Delta = r$ ($|X| = |E|$) corresponds to finite projective planes, which exist for $\Delta = q+1$ where q is a prime power: in that case $|X| = |E| = q^2+q+1$. However, for practical purposes, the interesting case is when Δ is small and therefore $\Delta < r$. By replacing, in a projective plane of order q (when it exists), each vertex j by k vertices $j_1, j_2, \ldots, j_k$ and each line with vertex set L by a bus with vertex set $B = \{j_i$ for $i = 1,2,\ldots,k \mid j \in L\}$, we obtain a hypergraph of degree $\Delta = q+1$, uniform size $k(q+1)$ with $k(q^2+q+1)$ vertices and diameter 1. We conjecture that we always have: $n(\Delta,D,1) \leqslant \left\lfloor \frac{(\Delta^2-\Delta+1)r}{\Delta} \right\rfloor$. We can prove this for $\Delta = 2$ and $\Delta = 3$, and in this case the bound is attained for an infinite number of values.

Dual hypergraphs

Let us recall that if $H = (X,E)$ is a hypergraph, its dual hypergraph $H^* = (E^*, X^*)$ is the hypergraph whose vertices (resp. edges) are in one-to-one correspondence with the edges (resp. vertices) of H and where the vertex e_j^* belongs to the edge X_i^* if and only if x_i belongs to E_j. If H is a (Δ,D,r)-hypergraph with n vertices and m edges, then H^* is an (r,D^*,Δ)-hypergraph with m vertices and n edges. Though we know that we have the relation $D^* = D$ or $D^* = D+1$ or $D^* = D-1$ between the diameter D^* of H^* and the diameter D of H, the characterization or the construction of graphs satisfying one of those equalities seems to be a difficult problem, even if H is a graph (in this case D^* is equal to the diameter of the line-graph). However, if H is bipartite then $D^* \leqslant D$. But, here again the characterization of the graphs satisfying $D^* = D$ appears to be difficult. The dual version of the determination of $n(\Delta,1,r)$ can be reformulated as follows: find the maximum cardinality of a set system (intersecting family), that is a family of sets E_i such that $E_i \cap E_j \neq \emptyset$, $|E_i| \leqslant \Delta$ and every vertex belongs to at most r sets. Without the last condition, the problem is well known; the solutions are the Δ-systems. Apparently there is no article with the last restriction on the degree.

Case $\Delta = 2$

One important practical case is the determination of $(2,D,r)$-hypergraphs having a large number of vertices. Since the dual of a $(2,D,r)$-hypergraph is a graph, the problem becomes an extremal problem in graph theory. Let us define the distance between two edges e and e' of a graph as the minimum number of vertices in a path $e = e_1, x_1, e_2, x_2, \ldots, e_i, x_i, \ldots, e_p = e'$ joining e and e'. Let the line diameter of a graph be the maximum of the distances over all pairs of edges. The line diameter of a graph is therefore the diameter of its dual hypergraph and also the diameter of its line-graph. The determination of $n(2,D,r)$ is equivalent to that of the maximum number of edges of a graph G whose line diameter is D and whose maximum degree is r. Recall that if a graph is r-regular, finding the maximum number of edges is equivalent to finding the maximum number of vertices $(m = \frac{1}{2}nr)$.

As any graph of diameter $D-1$ and any bipartite graph of diameter D have line-diameter at most D, we can use either $(r,D-1)$-graphs or (r,D)-bipartite graphs to construct large $(2,D,r)$-hypergraphs.

For example, consider the case $D = 2$. We can use the best-known $(r,1)$-graph, which is trivially K_{r+1} and therefore we have a $(2,2,r)$-hypergraph with $\frac{1}{2}r(r+1)$ vertices. We can consider the best $(r,2)$-bipartite graph, which is $K_{r,r}$, and therefore we have a $(2,2,r)$-hypergraph with r^2 vertices (note that it is the grid or mesh considered in the introduction to this section). But in this case, the graph $C_5 \otimes S_t$, obtained from C_5 by replacing each vertex i $(i = 1, 2, 3, 4 \text{ or } 5)$ by t vertices and each edge by a complete bipartite graph $K_{t,t}$, has $5t$ vertices, $5t^2$ edges and degree $2t$. We can check that it has line diameter 2 , and therefore its dual also has diameter 2 , so it is a $(2,2,r)$-hypergraph on $\frac{5}{4}r^2$ vertices with $r = 2t$. Therefore $n(2,2,r) \geq \frac{5}{4}r^2$ if r is even. Trotter (private communication, June 1982) proved that $n(2,r,r) \leq \frac{3}{2}r^2$, and very recently, answering one of our conjectures, it has been shown by Kleitman (1983) that every graph of maximum degree r and line diameter 2 has at most $\frac{5}{4}r^2$ vertices. Thus $n(2,2,r) \leq \frac{5}{4}r^2$ and if r is even $n(2,2,r) = \frac{5}{4}r^2$ (compare this bound with the Moore bound $2r^2-2r+1$) . In the general case, we do not know the exact answer. Let v_D be $\liminf_{r\to\infty} n(2,D,r)r^{-D}$; by Benson (1966) and Delorme (1983b)'s results on bipartite graphs, we have: $v_3 \geq 1$, $v_4 \geq 1$, $v_6 \geq 1$ and in general $v_D \geq (\frac{1}{2})^{D-1}$. (Note that this result improved the bound proposed by Finkel & Solomon (1980).)

General constructions

Most of the known constructions for graphs can be extended to hypergraphs. For example, we can use the compounding technique developed in Bermond, Delorme & Quisquater (1982b) by using BIBD instead of complete graphs. If we replace each vertex of an $(n(r-1)+1,r,1)$-BIBD by a (Δ,D,r)-hypergraph on n vertices, we obtain a $(\Delta+1,2D+1,r)$-hypergraph on $n^2(r-1) + n$ vertices. Therefore, by using $n(2,1,r) = \frac{3}{2}r$ if r is even (see case $D = 1$) , we obtain $n(3,3,r) \geq \frac{9r^3-9n^2+6r}{4}$ if r is even. Such constructions are described in Bermond, Bond & Saclé (1983). One can also use the hypergraph structure, for example by taking r copies of a (Δ,D,r)-hypergraph and joining by an edge the vertices having the same label. Other special constructions have also been developed by Quisquater (private communication, November 1982).

Here we have considered only the generalization of the (Δ,D)-graph problem; but it would be worth extending the problems that we have studied in the previous sections for graphs to the case of hypergraphs.

Acknowledgements

We wish to thank all those who helped us with the preparation of this survey, in particular J.A. BONDY, C. DELORME, J.J. QUISQUATER and U. SCHUMACHER.

REFERENCES

Amar, D. (1983). On the connectivity of some telecommunication networks. IEEE Trans. Comput. To appear.

Amar, D., Fournier, I. & Germa, A. (1983). Ordre minimum d'un graphe simple de diamètre degré minimum et connexité donnés. To appear.

Arden, B.W. & Lee, H. (1982). A regular network for multicomputer systems. IEEE Trans. Comput., C-31, 60-69.

Aschbacher, M. (1971). The nonexistence of rank three permutation groups of degree 3250 and subdegree 57. J. Algebra, 19, 538-540.

Ball, M.O. (1980). Complexity of network reliability computations. Networks, 10, 153-165.

Benson, C.T. (1966). Minimal regular graphs of girths eight and twelve. Canad. J. Math., 18, 1091-1094.

Berge, C. (1973). Graphs and Hypergraphs. North-Holland, Amsterdam.

Bermond, J.C. & Bollobás, B. (1981). The diameter of graphs - a survey. In Proc. 12th S.E. Conf. Congressus Numerantium, 32, 3-27.

Bermond, J.C., Bond, J. & Saclé, J.F. (1983). Large hypergraphs with given degree, diameter and edge size. Article in preparation.

Bermond, J.C., Delorme, C. & Farhi, G. (1982). Large graphs with given degree and diameter III. In Proc. Coll. Cambridge 1981. Ann. Discrete Math., 13, 23-32.

Bermond, J.C., Delorme, C. & Farhi, G. (1983). Large graphs with given degree and diameter II. Submitted to J. Combin. Theory.

Bermond, J.C., Delorme, C. & Quisquater, J.J. (1982a). Tables of large graphs with given degree and diameter. Inform. Process. Lett., 15, 10-13.

Bermond, J.C., Delorme, C. & Quisquater, J.J. (1982b). Grands graphes de degrés et diamètre fixés. In Proc. Coll. CNRS, Marseille, 1981. Ann. Discrete Math., 17, 65-73.

Biggs, N. (1974). Algebraic Graph Theory. Cambridge Tracts in Math., 67. Cambridge Univ. Press, London.

Bloom, G.S., Kennedy, J.W. & Quintas, L.V. (1981). Distance degree regular graphs. In The Theory and Applications of Graphs (Proc. Fourth International Conference, Kalamazoo, 1980), Wiley, New York, 95-108.

Boesch, F.T. (1983). Graph theoretic models for network reliability studies. Technical Report 8010, Electrical Engineering and Computer Science Department, Stevens Inst. Tech., Hoboken, New Jersey.

Boesch, F.T., Harary, F. & Kabell, J.A. (1981). Graphs as models of communication network vulnerability: connectivity and persistence. Networks, 11, 57-63.

Bollobás, B. (1971). Graphs with given diameter and maximal valency and with a minimal number of edges. In Combinatorial Mathematics and its Applications, Academic Press, London, 25-37.

Bollobás, B. (1978). Extremal Graph Theory. Academic Press, London.

Bollobás, B. (1981). The diameter of random graphs. Trans. Amer. Math. Soc., 267, 41-52.

Bollobás, B. & Eldridge, S. (1976). On graphs with diameter 2. J. Combin. Theory Ser. B, 21, 201-205.

Bollobás, B. & de la Vega, W.F. (1983). The diameter of random graphs. To appear.

Bond, J., Moulinoux, V. & Peyrat, C. (1983). On large graphs with given diameter, degree and vulnerability. Article in preparation.

Bondy, J.A. & Hell, P. (1983). Counterexamples to theorems of Menger type for the diameter. Discrete Math., 44.

Bondy, J.A. & Murty, U.S.R. (1976). Graph Theory with Applications. American Elsevier, New York.

Bosák, J. (1978). Geodetic graphs. In Combinatorics (Proc. Fifth Hungarian Colloq., Keszthely, 1976), Vol I, North-Holland, Amsterdam, 151-172.

Bosák, J., Kotzig, A. & Znám, S. (1968). Strongly geodetic graphs. J. Combin. Theory, 5, 170-176.

Bose, R.C. & Dowling, T.A. (1971). A generalization of Moore graphs of diameter two. J. Combin. Theory, 11, 213-226.

Boyles, S.M. & Exoo, G. (1982). A counterexample to a conjecture on paths of bounded length. J. Graph Theory, 6, 205-209.

Bridges, W.G. & Toueg, S. (1980). On the impossibility of directed Moore graphs. J. Combin. Theory Ser. B, 29, 339-341.

de Bruijn, N.G. (1946). A combinatorial problem. Nederl. Akad. Wetensh. Proc., 49, 758-764 = Indag. Math., 8, 461-467.

Buckley, F. (1981). Mean distance in line graphs. In Proc. Twelfth S.E. Conf., Baton Rouge. Congressus Numerantium, 32, 153-162.

Buckley, F. & Superville, F. (1981). Distance distributions and mean distance problems. In Proc. Third Caribbean Conf. Combinatorics & Computing. University of West Indies, Cave Hill, 67-76.

Buekenhout, F. (1983). (g,d^*,d)-gons. In Proc. Conference in Honour of T.G. Ostrom. To appear.

Caccetta, L. (1979). On extremal graphs with given diameter and connectivity. In Topics in Graph Theory (Proc. New York Acad. Sci. Scientist in-Residence Program, New York 1977). Ann. New York Acad. Sci., 328, 76-94.

Caccetta, L. & Häggkvist, R. (1979). On diameter critical graphs. Discrete Math., 28, 223-229.

Cerf, V.G., Cowan, D.D., Mullin, R.C. & Stanton, R.G. (1974a). Computer networks and generalized Moore graphs. In Proc. Third Manitoba Conference on Numerical Mathematics and Computing (Winnipeg, Man., 1973). Congressus Numerantium, 9, 379-398.

Cerf, V.G., Cowan, D.D., Mullin, R.C. & Stanton, R.G. (1974b). A partial census of trivalent generalized Moore networks. In Combinatorial Mathematics III (Proc. Third Australian Conf., Univ. Queensland, St. Lucia, 1974). Springer Lecture Notes in Math., 452, 1-27.

Cerf, V.G., Cowan, D.D., Mullin, R.C. & Stanton, R.G. (1975). Topological design considerations in computer communication networks. In Computer Communication Networks. NATO Adv. Study Inst. Ser. E: Appl. Sci., 4, 101-112.

Cerf, V.G., Cowan, D.D., Mullin, R.C. & Stanton, R.G. (1976). Some unique extremal graphs. Ars Combin., 1, 119-157.

Christofides, N. & Whitlock, C.A. (1981). Network synthesis with connectivity constraints; a survey. In Operational Research '81 (Proc. IFORS International Conference on Operational Research, Hamburg), North-Holland, Amsterdam, 655-673.

Chung, F.R.K. (1983). Problem. In Proc. Sixth Hungarian Colloquium on Combinatorics (Eger, 1981), Janos Bolyai Math. Soc. To appear.

Chung, F.R.K. & Garey, M.R. (1983). Diameter bounds for altered graphs. To appear.

Chvátal, V. (1983). Linear Programming. W.H. Freeman & Co., San Francisco. To appear.

Damerell, R.M. & Georgiacodis, M.A. (1981a). On Moore geometries I. J. London Math. Soc. (2), 23, 1-9.

Damerell, R.M. & Georgiacodis, M.A. (1981b). On Moore geometries II. Math. Proc. Cambridge Philos. Soc., 90, 33-40.

Delorme, C. (1983a). Grands graphes de degré et diamètre donné. Submitted to European J. Combin.

Delorme, C. (1983b). Large bipartite graphs with given degree and diameter. To appear.

Delorme, C. (1983c). Régularité métrique forte. To appear.

Delorme, C. & Farhi, G. (1983). Large graphs with given degree and diameter I. Submitted to IEEE Trans. Comput.

Doty, K.W. (1982). Large regular interconnection networks. Proc. International Conf. on Distributed Computing Systems (Miami, 1982), 312-317.

Doyen, J. & Rosa, H. (1980). Updated bibliography and survey of Steiner systems. Ann. Discrete Math., 7, 317-349.

Doyle, J.K. and Graver, J.E. (1977). Mean distance in a graph. Discrete Math., 17, 147-154.

Elspas, B. (1964). Topological constraints on interconnection limited logic. In Switching Circuit Theory and Logical Design, 5, 133-147.

Enomoto, H. & Usami, Y. (1983). Minimum number of edges in graphs with given diameter and connectivity I. Submitted to J. Combin. Theory Ser. B.

Entringer, R.C. & Jackson, D.E. (1982). Minimum k-geodetically connected graphs. In Proc. 13th S.E. Conf., Utilitas Math., 303-309.

Entringer, R.C., Jackson, D.E. & Slater, P.J. (1977). Geodetic connectivity of graphs. IEEE Trans. Circuits and Systems, CAS-24, 460-463.

Erdös, P. Fajtlowicz, S. & Hoffman, A.J. (1980). Maximum degree in graphs of diameter 2. Networks, 10, 87-90.

Erdös, P. & Rényi, A. (1962). On a problem in the theory of graphs. Magyar Tud. Akad. Mat. Kutató Int. Közl., 7, 623-641.

Erdös, P., Rényi, A. & Sós, V.T. (1966). On a problem of graph theory. Studia Sci. Math. Hungar., 1, 215-235.

Even, S. (1979). Graph Algorithms. Computer Science Press, Woodland Hills, Calif.

Exoo, G. (1982). On a measure of communication network vulnerability. Networks, 12, 405-409.

Farley, A.M. & Hedetniemi, S.T. (1978). Broadcasting in grid graphs. In Proc. Ninth S.E. Conf. on Combinatorics, Graph Theory and Computing (Boca Raton, Flo.). Congressus Numerantium, 21, 275-288.

Feng, T.Y. (1981). A survey of interconnection networks. Computer, 14, 12-27.

Finkel, R.A. & Solomon, M.H. (1980). Processor interconnection strategies. IEEE Trans. Comput., C-29, 360-371.

Finkel, R.A. & Solomon, M.H. (1981). The lens interconnection strategy. IEEE Trans. Comput., C-30, 960-965.

Fiol, M.A., Alegre, I. & Yebra, J.L.A. (1983). Line digraph iteration and the (d,k) problem for directed graphs. Preprint.

Frank, H. & Frisch, I.T. (1970). Analysis and design of survivable networks. IEEE Trans. Comm. Tech., COM-18, 501-519.

Frank. H. & Frisch, I.T. (1971). Communication Transmission and Transportation Networks. Addison-Wesley, Reading, Mass.

Fuglister, F.J. (1977). On finite Moore geometries. J. Combin. Theory Ser. A, 23, 187-197.

Gewirtz, A. (1969). Graphs with maximal even girth. Canad. J. Math., 21, 915-934.

Glivjak, F. (1975). On the impossibility to construct diametrically critical graphs by extensions. Arch. Math. (Brno), 11, 131-137.

Glivjak, F. (1976). A new proof of one estimation. Instituto Lombardo (Rend. Sci.), A 110, 3-5.

Golunkov, Y.V. (1973). On complexity of representing permutations of symmetric semigroups in terms of the elements of the system of generators. Kibernetika (Kiev), (1971), No. 1, 43-44. Translation in Cybernetics, 56-58.

Hanani, H. (1975). Balanced incomplete block designs and related designs. Discrete Math., 11, 255-369.

Harary, F. (1962). The maximum connectivity of a graph. Proc. Nat. Acad. Sci. USA, 48, 1142-1146.

Hartman, J. & Rubin, I. (1978). On diameter stability of graphs. In Theory and Applications of Graphs. Springer Lecture Notes in Math., 642, 247-254.

Hirschberg, D.S. & Wong, C.K. (1979). Upper and lower bounds for graph-diameter problems with application to record allocation. J. Combin. Theory Ser. B, 26, 66-74.

Hoffman, A.J. & Singleton, R.R. (1960). On Moore graphs with diameters 2 and 3. IBM J. Res. Develop., 4, 497-504.

Hwang, F.K. (1978). Superior channel graphs. Coll. ITC.

Imase, M. & Itoh, M. (1981). Design to minimize diameter on building-block network. IEEE Trans. Comput., C-30, 439-442.

Itai, A., Perl, Y. & Shiloach, Y. (1982). The complexity of finding maximum disjoint paths with length constraints. Networks, 12, 277-286.

Jerrum. M. & Skyum, S. (1983). Families of fixed degree graphs for processor interconnection. Internat. Report, Depart. of Comput. Science, Univ. Edinburgh.

Kantor, W.M. (1977). Moore geometries and rank 3 groups having $\mu = 1$. Quart. J. Math. Oxford Ser. (2), 28, 309-328.

Kautz, W.H. (1969). The design of optimum interconnection networks for multiprocessors.

Klee, V. & Larman, D. (1981). Diameters of random graphs. Canad. J. Math., 33, 618-640.

Kleitman, D.J. (1983). Private communication of Trotter, W.T.

Krishnamoorthy, V. & Nandakumar, R. (1981). A class of counterexamples to a conjecture on diameter critical graphs. In Combinatorics and Graph Theory. (Proc. Second Symp. Calcutta). Springer Lecture Notes in Math. 885, 297-300.

Kuich, W. & Sauer, N. (1969). On the existence of certain minimal regular n-systems with given girth. In Proof Techniques in Graph Theory (Proc. Second Ann Arbor Graph Theory Conf.), Academic Press, New York, 93-101.

Lam, C.W.H. & van Lint, J.H. (1978). Directed graphs with unique paths of fixed length, J. Combin. Theory Ser. B, 24, 331-337.

Leland, W. E. & Solomon, M.H. (1982). Dense trivalent graphs for processor interconnection. IEEE Trans. Comput., C-31, 219-222.

Leland, W., Finkel, R., Qiao, L., Solomon,M.H.& Uhr, L. (1981). High density graphs for processor interconnection. Inform. Process. Lett., 12, 117-120.

Lovász, L., Neumann-Lara, V. & Plummer, M. (1978). Mengerian theorems for paths of bounded length. Period. Math. Hungar., 9, 269-276.

Mader, W. (1970). Über den Zusammenhang symmetrischer Graphen. Arch. Math. (Basel), 21, 331-336.

Mader, W. (1971). Minimale n-fach kantenzusammenhängende Graphen. Math. Ann., 191, 21-28.

Mader, W. (1979). Connectivity and edge-connectivity in finite graphs. In Surveys in Combinatorics (Proc. Seventh British Combinatorial Conference, Cambridge, 1979). London Math. Soc. Lecture Note Series, 38, 66-95.

Malhotra, W.M., Kumar, M.P. & Maheshwari, S.N. (1978). An $O(|V|^3)$ algorithm for finding maximum flows in networks. Inform. Process. Lett. 7, 277-278.

Memmi, G. & Raillard, Y. (1982). Some new results about the (d,k)-graph problem. IEEE Trans. Comput. C-31, 784-791.

Murty, U.S.R. & Vijayan, K. (1964). On accessibility in graphs. Sankhyā Ser. A, 26, 299-302.

Myers, B.R. (1981). Regular separable graphs of minimum order with given diameter. Discrete Math., 33, 289-311.

Ott, U. (1983). Bericht über Hecke Algebren und Coxeter Algebren endlicher Geometrien. In Proc.Conf.Finite Geometries. Sussex 1980. To appear.

Pach, J. & Surányi, L. (1981). Graphs of diameter 2 and linear programming. In Proc. Colloq. Algebraic Methods in Graph Theory. Coll. Math. Soc. János Bolyai, 25, vol. 2. North-Holland, Amsterdam, 599-629.

Paoli, M., Wong, C.K. & Wong, W.W. (1983). Minimum k-hamiltonian graphs. Article in preparation.

Plesník, J. (1975a). Note on diametrically critical graphs. In Recent Advances in Graph Theory (Proc. Second Czechoslovak Sympos., Prague, 1974), Academia, Prague, 455-465.

Plesník, J. (1975b). Critical graphs of given diameter. Acta Fac. Rerum Natur. Univ. Comenian. Math., 30, 71-93.

Plesník, J. (1981). The complexity of designing a network with minimum diameter. Networks, 11, 77-85.

Plesník, J. & Znám. (1974). Strongly geodetic directed graphs, Acta Fac. Rerum Natur. Univ. Comenian. Math., 29, 29-34.

Quaife, H.J.(1969). On (d,k,μ) graphs. IEEE Trans. Comput., C-18, 270-272.

Quisquater, J.J. (1983a). New constructions of large graphs with fixed degree and diameter. To appear.

Quisquater, J.J. (1983b). Diameter of Cayley graphs of symmetric groups. To appear.

Raynal, D. (1983). Ordre minimum d'un hypergraphe h-uniforme, de degré minimum et de diamètre donnés. Rapport de stage LRI.

Roos, C. & van Zanten, A.J. (1983a). On the existence of certain generalized Moore geometries I. To appear.

Roos, C. & van Zanten, A.J. (1983b). On the existence of certain generalized Moore geometries II. To appear.

Roos, C. & van Zanten, A.J. (1983c). On a class of distance regular graphs. To appear.

Schumacher, U. (1983). An algorithm for construction of a k-connected graph with minimum number of edges and quasiminimal diameter. Networks, to appear.

Singleton, R.R. (1966). On minimal graphs of maximum even girth. J. Combin. Theory, _1_, 306-332.

Stemple, J.G. (1979). Geodetic graphs homeomorphic to a complete graph. Ann. New York Acad. Sci., _319_, 512-517.

Storwick, R.M. (1970). Improved construction techniques for (d,k) graphs. IEEE Trans. Comput. C-_19_, 1214-1216.

Suvikov (1978). The minimal number of edges in a doubly connected graph with given diameter. Principles of the Construction of Information Disseminating Devices, Nauka, Moscow, 87-90.

Tansel, B.C., Francis, R.L. & Lowe, T.J. (1983). Location on networks: a survey. Research Report, Industrial Systems Engineering Dept., Univ. Florida, Gainesville.

Tchuente, M. (1982). Parallel realization of permutations over trees. Discrete Math., _39_, 211-214.

Toueg, S. & Steiglitz, K. (1979). The design of small-diameter networks by local search. IEEE Trans. Comput., C-_28_, 537-542.

Uhr, L. (1983). Compounding denser (d,k) graph architectures for computer networks. Comput. Sci. Tech. Rept. N 414, Univ. Wisconsin, Madison.

Watkins, M.E. (1970). Connectivity of transitive graphs. J. Combin. Theory, _8_, 23-29.

Wegner, G. (1976). Graphs with given diameter and a colouring problem. Unpublished paper.

Wilkov, R.S. (1972). Analysis and design of reliable computer networks. IEEE Trans. Comm., COM-_20_, 660-678.

Wittie, L.D. (1981). Communication structures for large networks of microcomputers. IEEE Trans. Comput., C-_30_, 264-273.

Wittie, L.D. & Van Tilborg, A.M. (1980). MICROS, a distributed operating system for MICRONET, a reconfigurable network computer. IEEE Trans. Comput., C-_29_, 1133-1144.

Wong, P.K. (1982). Cages - a survey. J. Graph Theory, _6_, 1-22.

Yannakakis, M. (1978). Node- and edge-deletion NP-complete problems. _In_ Proc. Tenth Annual ACM Symp. Theory of Computing, ACM, New York, 253-264.

The Friendship Theorem and the Love Problem

By J. M. Hammersley

Trinity College, Oxford

> "The nymph Echo fell in love with Narcissus, but was repulsed. Aphrodite punished him for his cruelty by making him enamoured of his own image in a fountain. His fruitless attempts to approach this beautiful object led to his despair and death. He was changed into the flower that bears his name."— Sir Paul Harvey, *Oxford Companion to Classical Literature*.

1. Community relations.

In response to the editor's request for a survey article on a combinatorial topic, I have assembled under a single heading —the love problem, as I shall call it — material that has previously appeared in separate contexts and diverse guises, such as the solubility of Diophantine quadratic matrix equations, the construction of block designs, the existence of finite geometries, etc. However, I shall only mention a handful of references as leads into the extensive literature, which I could not hope to cover by anything approaching a complete bibliography. Moreover, in stressing the graph-theoretic aspects of the matter, I shall be adopting a rather different line from traditional treatments.

Let us begin with a special case of the love problem, known as the friendship theorem. I do not know who first stated this theorem: the earliest published paper that I have come across is Wilf (1971), who cites an earlier unpublished account by Graham Higman in 1968.

THEOREM 1. (Friendship) *Suppose that, in a finite community of n people, any two distinct individuals have exactly one mutual friend. Then n cannot be even, although any odd* $n \geqslant 3$ *is possible. Further, the people can be labelled* $V_1, V_2, \ldots, V_n$ *such that, whenever* $2 \leqslant 2r < n$, V_{2r} *and* V_{2r+1} *are friends of each other and of* V_1 *and of nobody else. Thus* V_1, *the Dale Carnegie* (1953) *of the community, is everybody's friend.*

By hypothesis, friendship is symmetric but not reflexive: if V_i is a friend of V_j, then V_j is a friend of V_i; but V_i cannot befriend himself. Love, on the other hand, may or may not be reciprocated, and may be narcissistic. The love problem is the generalization of Theorem 1 when the asymmetric and possibly reflexive relationship of love replaces the symmetric and non-reflexive relationship of friendship. Whereas the friendship theorem is completely solved, the love problem is largely unsolved.

We shall write $V_i \to V_j$ (or equivalently $V_j \leftarrow V_i$) to signify that V_i loves V_j. It is possible that both $V_i \to V_j$ and $V_j \to V_i$, in which case we write $V_i \rightleftarrows V_j$.

Let V_i and V_j be any two distinct individuals in the community, and consider the three possible statements:

$$S_1(i, j)\colon V_i \to V_k \to V_j \text{ has a unique solution } k = k_1(i, j);$$
$$S_2(i, j)\colon V_i \leftarrow V_k \to V_j \text{ has a unique solution } k = k_2(i, j);$$
$$S_3(i, j)\colon V_i \to V_k \leftarrow V_j \text{ has a unique solution } k = k_3(i, j).$$

Here the notation $k(i, j)$ indicates that the solutions may depend upon i and j. The three integers k_1, k_2, k_3 need not be equal to one another, and they may or may not be equal to i and j. Clearly S_1 is a statement about the ordered pair $i \neq j$, whereas S_2 and S_3 are statements about the unordered pair $i \neq j$. For a community of size n, we may now formulate four versions of the love problem:

- $\mathfrak{L}_n^1$: the ordered love problem, in which $S_1(i, j)$ is true for each and every ordered pair $i \neq j$;
- $\mathfrak{L}_n^2$: the unordered love problem, in which $S_2(i, j)$ is true for each and every unordered pair $i \neq j$;
- $\mathfrak{L}_n^3$: the dual unordered love problem, in which $S_3(i, j)$ is true for each and every unordered pair $i \neq j$;
- $\mathfrak{L}_n^4$: the self-dual love problem, in which both $S_2(i, j)$ and $S_3(i, j)$ are true for each and every unordered pair $i \neq j$.

We shall say that a solution of any of these versions has narcissism m if $V_i \rightleftarrows V_i$ for exactly m values of i. Going beyond classical mythology, in which Narcissus loved only himself, we shall not forbid an individual, who loves himself, from loving others as well.

Evidently, the friendship theorem is the particular case of $\mathfrak{L}_n^1$ under the further restrictions that $m = 0$ and $k_1(i, j) = k_1(j, i)$ for all $i \neq j$. Alternatively, it is the particular case of $\mathfrak{L}_n^4$ with $m = 0$ and $k_2(i, j) = k_3(i, j)$ for all $i \neq j$. The friendship theorem therefore always provides a solution for any version of the love problem for odd $n \geqslant 3$, but it is a rather trivial and dull solution. We shall be more interested in other more exciting solutions, in which love outruns friendship. We shall concentrate mainly upon $\mathfrak{L}_n^1$ and $\mathfrak{L}_n^4$. Both these versions have a family of rather trivial solutions (including the friendship theorem) as well as a family of more interesting solutions. The following remarks in this paragraph briefly give the main facts for these more interesting solutions. Each individual loves, and is loved by, exactly c individuals where c is the least integer such that $c \geqslant \sqrt{n}$. For $\mathfrak{L}_n^4$, a necessary (but not sufficient) condition for the existence of an interesting solution is $n = c^2 - c + 1$; and , if some solution exists for that n, then solutions also exist for every narcissism $0 \leqslant m \leqslant n$. For $\mathfrak{L}_n^1$, the conditions upon n are less stringent, but the conditions upon m are much more stringent: it is necessary that $c^2 - c + 1 \leqslant n \leqslant c^2$ and for certain values of n only one value of m is possible; sometimes more than one value of m can occur but in no case can m be as large as n.

2. Graphical representation.

We can represent a solution of $\mathfrak{L}_n^l$ ($l = 1, 2, 3, 4$) by a partially directed graph $\mathfrak{G}_n^l$ with vertices $V_1, V_2, ..., V_n$. Two vertices V_i and V_j are connected by an undirected edge if and only if $V_i \rightleftharpoons V_j$, or by a directed edge from V_i to V_j if and only if $V_i \to V_j$ but not $V_j \to V_i$. It will be convenient to have a couple of short words to distinguish these two sorts of edge. We reserve the word *arc* for a directed edge, and the word *bib* for an undirected edge; and, if V_iV_j is a directed edge, our convention will be that it is directed from its first letter V_i to its last letter V_j. This terminology is mnemonic in that the first letter of *arc* precedes its last letter alphabetically, whereas the words *bib* and *edge* do not exhibit such alphabetic precedence. We retain the word *edge* to mean indifferently either a bib or an arc. By the very formulation of the love problem, $\mathfrak{G}_n^l$ cannot have any multiple edges; but if the narcissism $m > 0$, it will have m loops. Two graphs are graphically equivalent if they only differ in the numbering of their vertices; and we shall only recognize the structure of a graph to within graphical equivalence.

We shall illustrate typical graphs by diagrams, in which the vertices are black dots, the bibs are double lines, and the arcs are single lines carrying an arrow to indicate their direction. For clarity, loops are not marked on the diagrams; but instead any black dot that stands for a narcissistic individual is enclosed in a small circle. Figure 1 illustrates the friendship solution for $n = 15$; whereas Figure 2 illustrates a solution of $\mathfrak{L}_{15}^1$ with $m = 0$. Comparison of these two diagrams confirms the common experience that there is more to love than friendship.

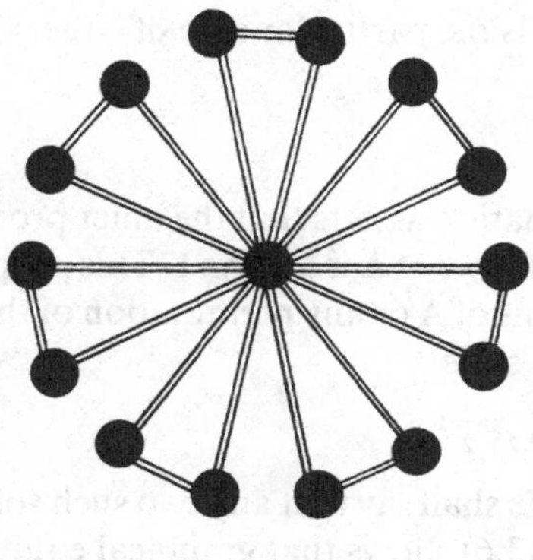

Figure 1.
(Friendship)
$\mathfrak{G}_{15}^1 = \mathfrak{G}_{15}^4$, $m = 0$.

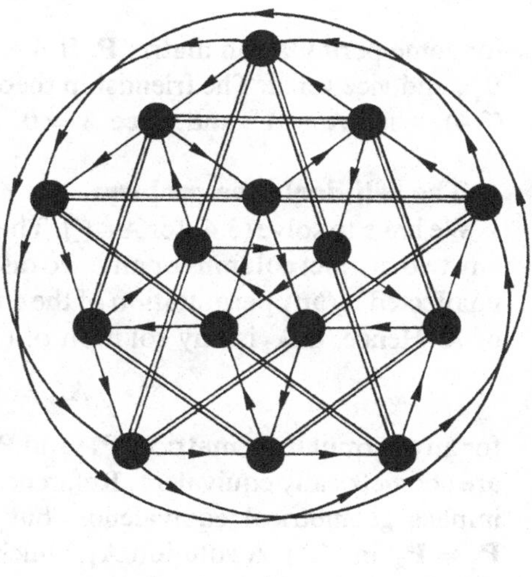

Figure 2. $\mathfrak{G}_{15}^1, m = 0$.

3. Algebraic representation.

We shall write $\square$ for the set of all $n \times n$ matrices whose elements can only assume the values 0 or 1. Particular members of $\square$ are the unit matrix (written **I**) and the matrix whose elements are all 1 (written **J**). A permutation matrix is a member of $\square$ with exactly one 1 in each row and each column. We write ∇ for the set of all $n \times n$ diagonal matrices: these are not necessarily members of $\square$. The transpose of a matrix **A** will be denoted by $\mathbf{A}^\mathsf{T}$. Any graph $\mathfrak{G}_n^l$ can be represented by its incidence matrix $\mathbf{A} \in \square$, whose elements $a_{ij} = 1$ or 0 according as $V_i \to V_j$ or not. For example, a bib connects V_i and V_j if and only if $a_{ij} = a_{ji} = 1$.

It is easy to see that the respective love problems amount to finding a solution $\mathbf{A} \in \square$ of the appropriate Diophantine quadratic matrix equations

$$\mathfrak{L}_n^1: \quad \mathbf{A}^2 - \mathbf{J} \in \nabla; \tag{3.1}$$

$$\mathfrak{L}_n^2: \quad \mathbf{A}^\mathsf{T}\mathbf{A} - \mathbf{J} \in \nabla; \tag{3.2}$$

$$\mathfrak{L}_n^3: \quad \mathbf{A}\mathbf{A}^\mathsf{T} - \mathbf{J} \in \nabla; \tag{3.3}$$

$$\mathfrak{L}_n^4: \quad \mathbf{A}^\mathsf{T}\mathbf{A} - \mathbf{J} \in \nabla, \ \mathbf{A}\mathbf{A}^\mathsf{T} - \mathbf{J} \in \nabla. \tag{3.4}$$

The narcissism of a solution is

$$m = \text{trace } \mathbf{A}. \tag{3.5}$$

Two solutions $\mathbf{A}_1$ and $\mathbf{A}_2$ are graphically equivalent if and only if

$$\mathbf{A}_1 = \mathbf{P}\mathbf{A}_2\mathbf{P}^{-1} \tag{3.6}$$

for some permutation matrix **P**. If **A** is a solution of $\mathfrak{L}_n^2$, then $\mathbf{A}^\mathsf{T}$ is a solution of $\mathfrak{L}_n^3$, and vice versa. The friendship theorem is the particular case of either (3.1) or (3.4), when $\mathbf{A} = \mathbf{A}^\mathsf{T}$ and trace $\mathbf{A} = 0$.

4. The self-dual love problem.

We have to solve (3.4) for $\mathbf{A} \in \square$. This equation asserts that the inner product of any two distinct columns or any two distinct rows of **A** is always 1. This property is unaffected by any permutation of the columns of **A** or any permutation of the rows of **A**. Hence, if $\mathbf{A}_2$ is any solution of (3.4), so is

$$\mathbf{A}_1 = \mathbf{P}_1\mathbf{A}_2\mathbf{P}_2^{-1} \tag{4.1}$$

for any permutation matrices $\mathbf{P}_1$ and $\mathbf{P}_2$. We shall say that any two such solutions are geometrically equivalent. Reference to (3.6) shows that graphical equivalence implies geometrical equivalence, but not conversely since we need not have $\mathbf{P}_1 = \mathbf{P}_2$ in (4.1). A solution $\mathbf{A}_1$, which is geometrically equivalent to $\mathbf{A}_2$, will be called an alias of $\mathbf{A}_1$; and a similar terminology will apply to the associated graphical representations $\mathfrak{G}_n^4$ of $\mathbf{A}_1$ and $\mathbf{A}_2$. To within graphical equivalence, we may arrange the rows of a solution **A** in any specified order (the columns then being rearranged in accordance with (3.6)); and then, keeping the order of the rows fixed, we generate the aliases of **A** by permuting the columns of **A**. However, depending upon the particular nature of **A**, not all of these permutations need lead to distinct aliases.

We illustrate this by considering a trivial solution of (3.4), namely

$$a_{i1} = a_{1j} = 1 \quad (i,j = 1,2,...,n); \quad a_{ij} = 0 \text{ otherwise.} \tag{4.2}$$

The $n!$ permutations of the columns of **A** lead only to 2 distinct aliases, depending upon whether or not the first column is shifted by the permutation. The associated graphs are illustrated in Figures 3 and 4 for $n = 15$, from which the graphs $\mathfrak{G}_n^4$ for general n will be readily perceived. The narcissism for (4.2) is always $m = 1$ or 2.

As a second illustration consider another trivial solution

$$a_{11} = 0; \quad a_{ii} = a_{i1} = a_{1j} = 1 \ (i,j = 2,...,n); \quad a_{ij} = 0 \text{ otherwise.} \tag{4.3}$$

If $n \geqslant 3$, then (4.3) is geometrically inequivalent to (4.2) and there are altogether $p(n-1)+p(n-2)$ distinct aliases of (4.3), where $p(n)$ denotes the number of partitions of n. To see this, consider the particular case $n = 15$ illustrated in Figures 5, 6, and 7. Figure 5 shows the graph of (4.3) without permutation of the columns. Figure 6 shows the graph that arises when the first column of **A** remains fixed, and the remaining columns undergo a permutation whose cycle decomposition has cycles of respective lengths 8, 4, 2. To get Figure 7, we interchange the first and last columns of **A** and subject the remaining 13 columns to a permutation whose cycle decomposition has cycles of length 5, 3, 2, 1, 1, 1. The friendship solution in Figure 1 would have been obtained by keeping the first column fixed and subjecting the remaining 14 columns to 7 disjoint transpositions. In general for any $n \geqslant 4$, we obtain an alias of (4.3) whose narcissism m has any prescribed value in $0 \leqslant m \leqslant n$. (If $n = 3$, there are aliases of (4.3) with $m = 0, 2, 3$ but not with $m = 1$).

The particular solutions (4.2) and (4.3) are rather special. To see this we now identify the rows of **A** with a set of points $P_1, P_2, ..., P_n$, and the columns of **A** with a set of lines $L_1, L_2, ..., L_n$. Each line is simply a subset of the set of n points, with the rule that $P_i \in L_j$ if and only if $a_{ij} = 1$. If $P_i \in L_j$, we say that P_i lies on L_j (or equivalently that L_j passes through P_i). The notion of geometrical equivalence, defined in (4.1), amounts to saying that the labels accorded to the points and to the lines are immaterial, and that we are only interested in the geometrical structure of the incidence relations. For example, to within geometrical equivalence, (4.2) is a geometry in which all points lie on one particular line and all lines pass through one particular point. Again, (4.3) represents a geometry with a pencil of $n-1$ lines through a particular point (say P_1), and a particular line (say L_1) that does not pass through P_1: the remaining $n-1$ points are the intersections of L_1 with lines going through P_1.

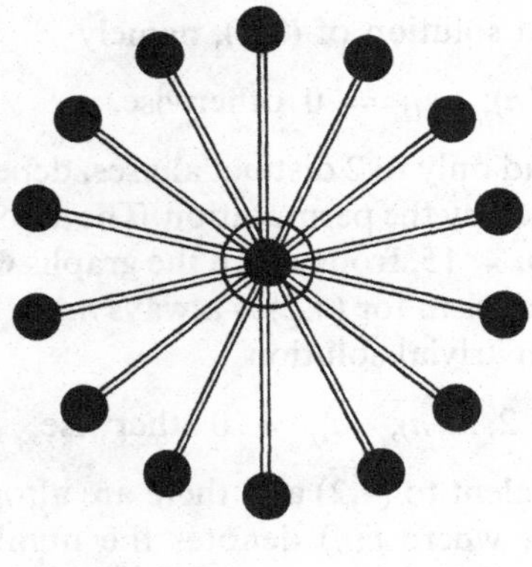

Figure 3. $\mathfrak{G}^4_{15}, m = 1$; (4.2).

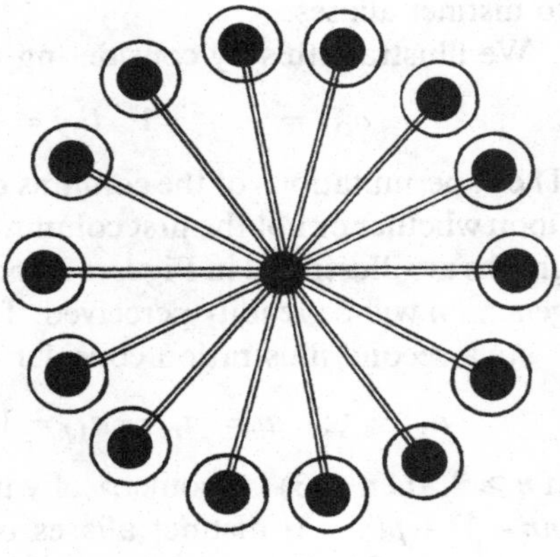

Figure 5. $\mathfrak{G}^4_{15}, m = 14$; (4.3).

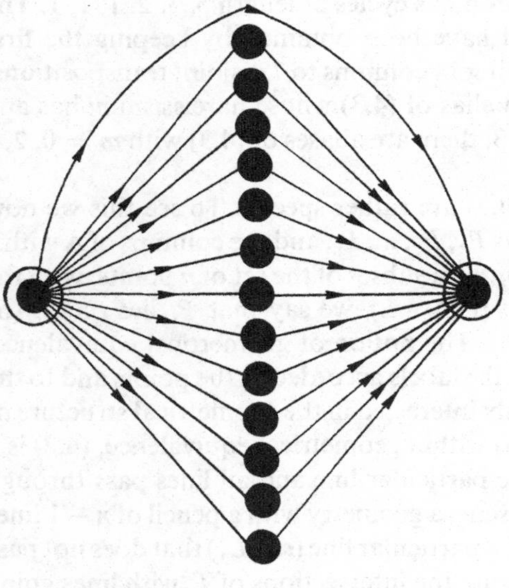

Figure 4. $\mathfrak{G}^4_{15}, m = 2$; (4.2).

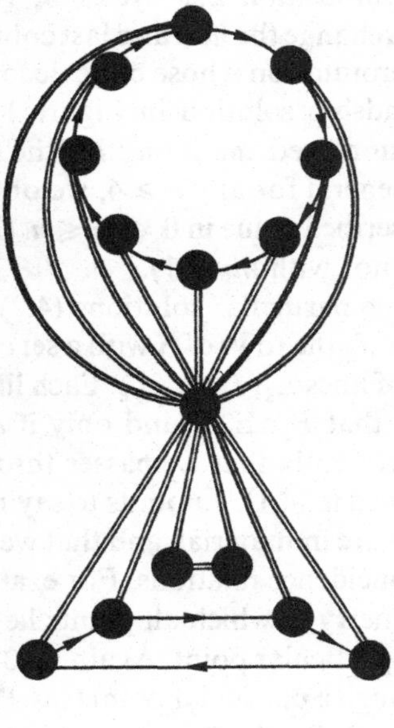

Figure 6. $\mathfrak{G}^4_{15}, m = 0$; (4.3).

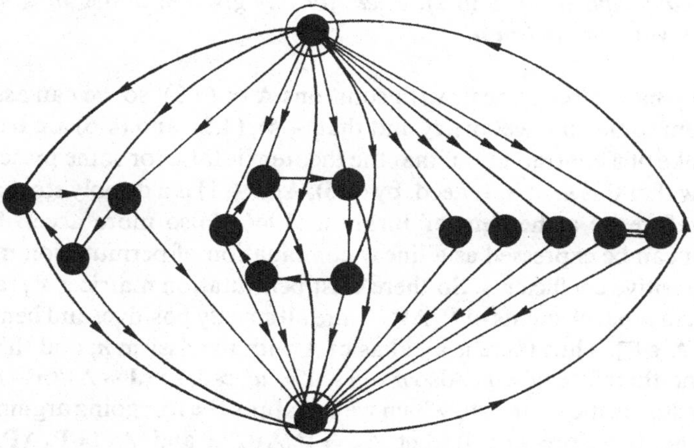

Figure 7. $\mathfrak{G}^4_{15}, m = 2$; (4.3).

Therefore, to get any other possible solution apart from (4.2) and (4.3), we must assume that the geometry contains at least 4 points, no 3 of which are collinear. We shall then have the following assertions as a result of (3.4):

(i) two distinct lines meet in a unique point;
(ii) two distinct points lie on a unique line;
(iii) there exist 4 points, of which no 3 are collinear.

These three assertions are, however, the axioms of a non-trivial finite projective geometry on n points ($n \geqslant 4$), from which it follows that

$$n = u^2 + u + 1 \tag{4.4}$$

for some integer $u \geqslant 2$, that each line passes through exactly $u + 1$ points, and that each point lies on exactly $u + 1$ lines. This means that the row and column sums of $\mathbf{A}$ are all $u + 1$. Thus

$$\mathbf{AJ} = \mathbf{JA} = (u+1)\mathbf{J}; \tag{4.5}$$

and (3.4) becomes

$$\mathbf{A}^{\mathsf{T}}\mathbf{A} = \mathbf{AA}^{\mathsf{T}} = u\mathbf{I} + \mathbf{J}. \tag{4.6}$$

The proof of (4.4) and the assertions which follow it may be found in Hall (1959) and Ryser (1963). These references also amplify the brief summary of the properties of finite geometries given in §5. It should be stressed at this stage that (4.4) is a necessary but not sufficient condition upon n: not all values of $u \geqslant 2$ lead to finite geometries.

THEOREM 2. *Let* $\mathbf{A}$ *be an* $n \times n$ *matrix satisfying* (3.4) *for* $n \geqslant 4$, *such that* $\mathbf{A}$ *is not geometrically equivalent to* (4.2). *Then, for any given* m *in* $0 \leqslant m \leqslant n$, *there is an alias of* $\mathbf{A}$ *with narcissism* m.

Proof: We have already dealt with solutions $\mathbf{A}$ of (4.3); so we can assume that $\mathbf{A}$ arises from some finite geometry and that (4.4), (4.5), and (4.6) are true. Suppose, for the sake of a contradiction, that the theorem is false for some prescribed m. We first show that this $m < n$. Indeed, by (4.5), $\mathbf{A}/(u+1)$ is a doubly stochastic matrix, and therefore by a theorem of Birkhoff (1946) (also more accessibly in Horn (1954)) it can be expressed as a linear combination of permutation matrices with strictly positive coefficients. So there exist permutation matrices $\mathbf{P}_1$ and $\mathbf{P}_2$ such that the diagonal elements of $\mathbf{P}_1\mathbf{A}\mathbf{P}_2^{-1}$ are all strictly positive, and hence equal to 1 because $\mathbf{A} \in \square$. Thus there is an alias of $\mathbf{A}$ with narcissism n; and this establishes $m \neq n$ and therefore $m < n$. Also $n - u - 1 = u^2 > 0$; so $(\mathbf{J} - \mathbf{A})/(n-u-1)$ is also a doubly stochastic matrix, to which we can apply the foregoing argument to prove $m \neq 0$ and therefore $m > 0$. Let $\mathbf{A}_1 = \mathbf{P}_3\mathbf{A}\mathbf{P}_4^{-1}$ and $\mathbf{A}_2 = \mathbf{P}_5\mathbf{A}\mathbf{P}_6^{-1}$ be these aliases of $\mathbf{A}$ with respective narcissisms 0 and n. By permuting the rows and columns of $\mathbf{A}_1$ we can convert $\mathbf{A}_1$ into $\mathbf{A}_2$. Let this permutation be performed by a sequence of transpositions of two rows or two columns each. Each transposition in the sequence can change the narcissism by only 0 or ± 1 or ± 2. So in changing the narcissism from 0 to n, the narcissism at some stage must assume both the values $m-1$ and $m+1$ in order to skip the assumed non-existent narcissism m. Let $\mathbf{A}_3$ be the matrix that has narcissism $m+1$ during the course of this sequence; so we have $\mathbf{A}_3 = \mathbf{P}_7\mathbf{A}\mathbf{P}_8^{-1}$ for some permutation matrices $\mathbf{P}_7$ and $\mathbf{P}_8$. For any permutation matrix $\mathbf{P}$, the narcissisms of $\mathbf{A}_4 = \mathbf{P}\mathbf{A}_3\mathbf{P}^{-1}$ and $\mathbf{A}_3$ are both equal to $m+1$. We choose $\mathbf{P}$ so that the first $m+1$ elements on the diagonal of $\mathbf{A}_4$ are 1 and the remaining diagonal elements (if any) are 0. We write $\mathbf{A}_4$ in partitioned form

$$\mathbf{A}_4 = \begin{bmatrix} \mathbf{A}_{11} & \mathbf{A}_{12} \\ \mathbf{A}_{21} & \mathbf{A}_{22} \end{bmatrix}, \tag{4.7}$$

where $\mathbf{A}_{11}$ is an $(m+1) \times (m+1)$ matrix. If $m+1 = n$, the other submatrices in the partition do not arise . Also, because $m > 0$, $\mathbf{A}_{11}$ has at least two rows and two columns. Now consider the elements a_{ij} and a_{ji} in $\mathbf{A}_{11}$ in which $i \neq j$ and $1 \leqslant i \leqslant m+1$ and $1 \leqslant j \leqslant m+1$. We cannot have $a_{ij} = a_{ji} = 1$ because the inner product of the ith and jth rows of $\mathbf{A}_4$ would then be at least 2, since $a_{ii} = a_{jj} = 1$. Nor can we have $a_{ij} = 1$, $a_{ji} = 0$; for in that case we could obtain an alias of narcissism m by interchanging the ith and jth columns of $\mathbf{A}_4$. Nor can we have $a_{ij} = 0$, $a_{ji} = 1$; for then we could achieve an alias of narcissism m by interchanging the ith and jth rows of $\mathbf{A}_4$. Since $\mathbf{A}_4 \in \square$, all the off-diagonal elements of $\mathbf{A}_{11}$ must be 0. Now the row and column sums of $\mathbf{A}_4$ are all equal to $u+1$ according to (4.5). Hence $\mathbf{A}_{12}$ must have at least 2 columns, and $\mathbf{A}_{21}$ must have at least 2 rows because $u \geqslant 2$. Next consider an element a_{ij} in $\mathbf{A}_{12}$, where $1 \leqslant i \leqslant m+1 < j \leqslant n$, and let a_{ji} be the corresponding element in $\mathbf{A}_{21}$. We cannot have $a_{ij} = a_{ji} = 0$, because an interchange of the ith and jth columns of $\mathbf{A}_4$ would yield an alias of narcissism m. Hence $a_{ij}+a_{ji} \geqslant 1$. This holds for each pair of corresponding elements in $\mathbf{A}_{12}$ and $\mathbf{A}_{21}$. Hence, if s denotes the sum of all the elements in $\mathbf{A}_{12}$ plus the sum of all the elements in $\mathbf{A}_{21}$, we have

$$s \geqslant (m+1)(n-m-1). \tag{4.8}$$

On the other hand, because $\mathbf{A}_{11}$ is a unit matrix, each row of $\mathbf{A}_{12}$ has a row sum u, and each column of $\mathbf{A}_{21}$ has a column sum u. So

$$s = 2u(m+1). \tag{4.9}$$

We shall next deduce that $u = 2$. There are three cases to consider:

(i) some column of $\mathbf{A}_{12}$ has column sum $u+1$,

(ii) some row of $\mathbf{A}_{21}$ has row sum $u+1$,

(iii) each row of $\mathbf{A}_{21}$ and each column of $\mathbf{A}_{12}$ has a sum not exceeding u.

These cases are exhaustive because the row and column sums of $\mathbf{A}_4$ are all $u+1$. In case (i), suppose that the jth column of $\mathbf{A}_4$ has all its 1's within $\mathbf{A}_{12}$; so it has zeros throughout $\mathbf{A}_{22}$. Since $\mathbf{A}_{11}$ is a unit matrix, the inner product of the ith and jth columns of $\mathbf{A}_4$, where $1 \leqslant i \leqslant m+1$, is $a_{ii}a_{ij} = 1$ for $j > m+1$. Hence $a_{ij} = 1$ for all $i \leqslant m+1$ and the column sum of the jth column of $\mathbf{A}_4$ is $m+1 = u+1$. Inserting this into (4.8) and (4.9) we get

$$2u(u+1) = s \geqslant (u+1)(n-u-1) = u^2(u+1) \tag{4.10}$$

by (4.4). Hence $u \leqslant 2$. However $u \geqslant 2$ for any finite geometry. So $u = 2$ as required. In case (ii) an exactly similar argument applies to the rows of $\mathbf{A}_4$. There remains case (iii), for which we have

$$s \leqslant 2u(n-m-1). \tag{4.11}$$

From (4.4), (4.8), and (4.9) we find $m \geqslant u(u-1)$; and (4.4), (4.9), and (4.11) yield $2m \leqslant u^2+u-1$. Eliminating m from these two inequalities we consequently get $u^2-3u+1 \leqslant 0$; and hence $u \leqslant 2$ while $u \geqslant 2$ as before. This completes the proof that $u = 2$.

For $u = 2$, the finite geometry is known to be unique, and within geometrical equivalence its incidence matrix is

	L_1	L_2	L_3	L_4	L_5	L_6	L_7
P_1	1	0	0	0	0	1	1
P_2	0	1	0	0	1	0	1
P_3	0	0	1	0	1	1	0
P_4	1	1	1	0	0	0	0
P_5	1	0	0	1	1	0	0
P_6	0	1	0	1	0	1	0
P_7	0	0	1	1	0	0	1

The graphical representation of this matrix has narcissism $m = 6$. But we can obtain other values of m by rearranging the above columns in the following orders:

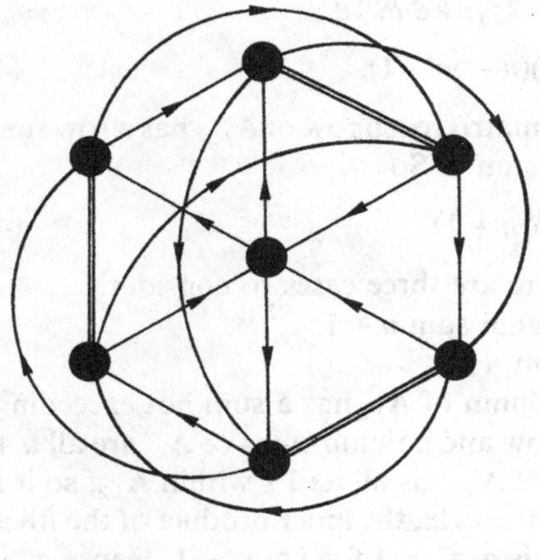

Figure 8. $\mathfrak{G}_7^4, m = 0$.

$m=0$: $L_2L_3L_4L_5L_6L_7L_1$
$m=1$: $L_2L_3L_4L_5L_1L_7L_6$
$m=2$: $L_2L_1L_3L_4L_6L_5L_7$
$m=3$: $L_2L_3L_4L_5L_1L_6L_7$
$m=4$: $L_2L_1L_3L_4L_5L_6L_7$
$m=5$: $L_7L_6L_2L_1L_5L_4L_3$
$m=6$: $L_1L_2L_3L_4L_5L_6L_7$
$m=7$: $L_7L_2L_6L_1L_5L_4L_3$.

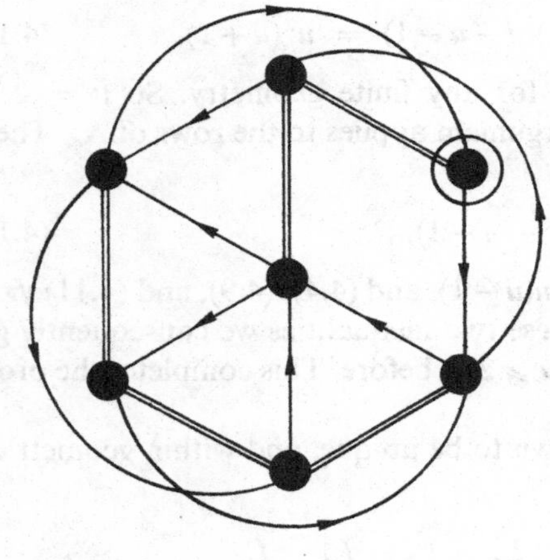

Figure 9. $\mathfrak{G}_7^4, m = 1$.

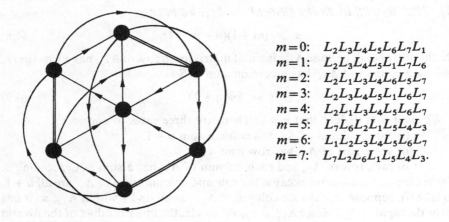

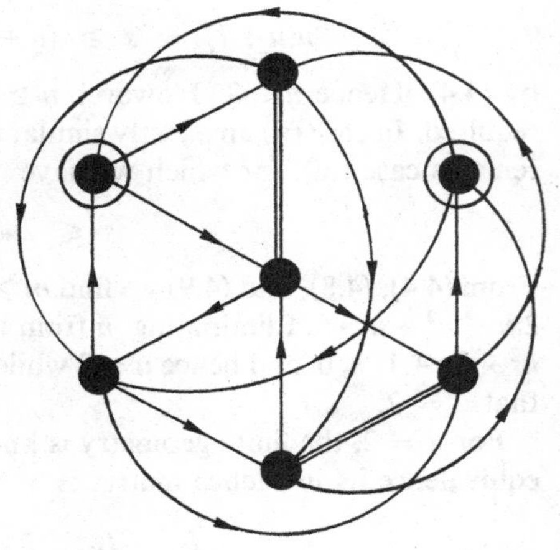

Figure 10. $\mathfrak{G}_7^4, m = 2$.

Thus all possible narcissisms occur even in the case $u = 2$. This contradiction completes the proof of Theorem 2.

Figures 8 to 15 provide the graphs for the foregoing aliases $m = 0, 1, ..., 7$ when $u = 2$. These graphs do not by any means exhaust all the aliases from this particular geometry with $u = 2$. For example, Figure 16 shows the alias that arises from taking the columns in the order $L_7L_6L_5L_4L_3L_2L_1$. Figure 16 is quite different in character from Figures 8 to 15: in Figure 16 love is always reciprocated

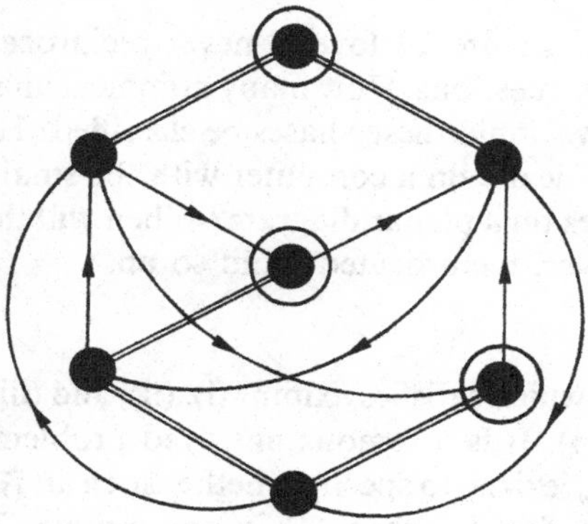

Figure 11. $\mathfrak{G}_7^4, m = 3$.

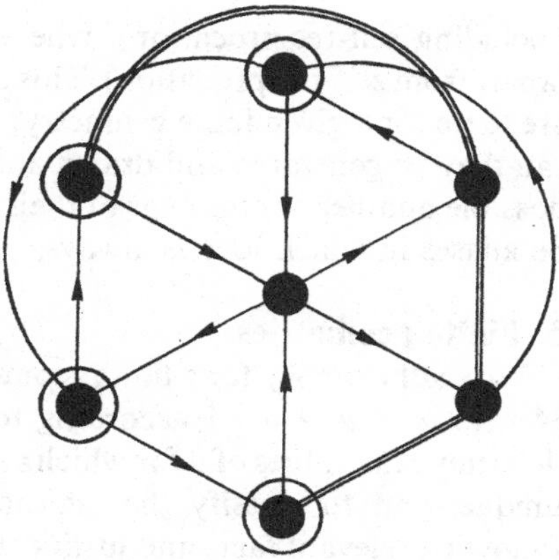

Figure 12. $\mathfrak{G}_7^4, m = 4$.

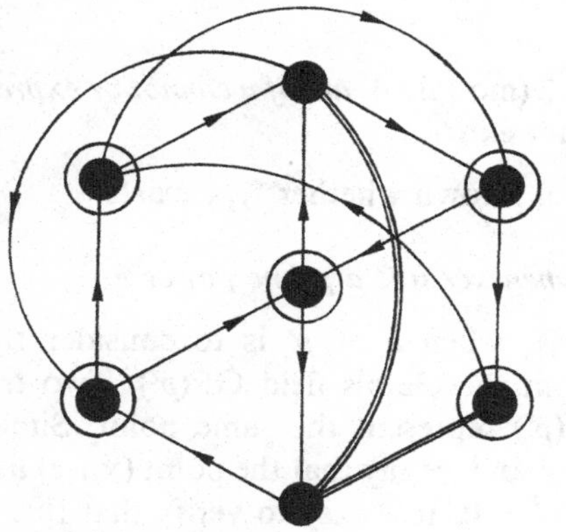

Figure 13. $\mathfrak{G}_7^4, m = 5$.

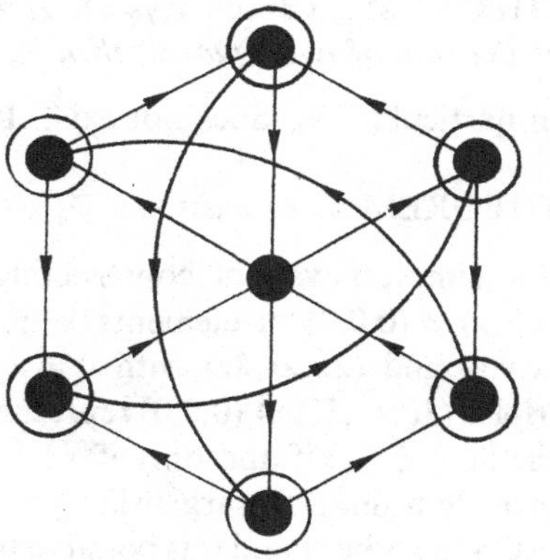

Figure 14. $\mathfrak{G}_7^4, m = 6$.

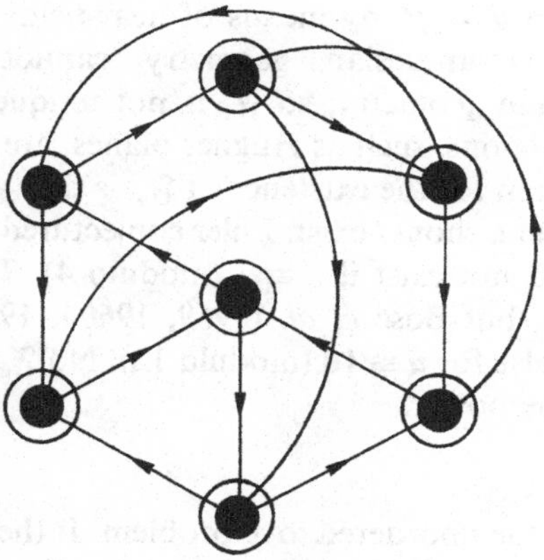

Figure 15. $\mathfrak{G}_7^4, m = 7$.

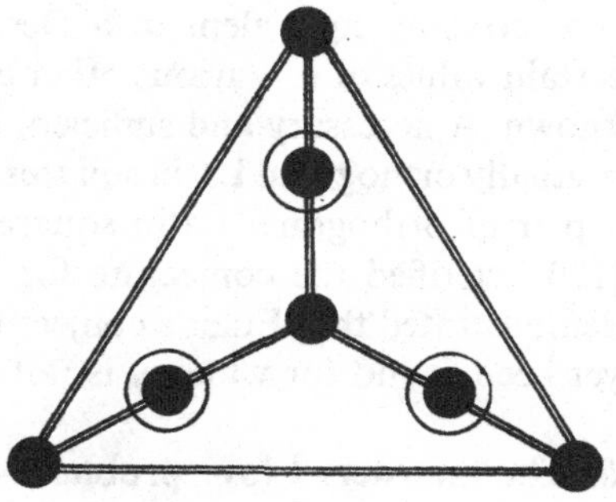

Figure 16. $\mathfrak{G}_7^1 = \mathfrak{G}_7^4, m = 3$.

(including self-reciprocation), whereas in Figure 14 love is never reciprocated (apart from self-reciprocation). This raises questions. How many graphical aliases are there for a given finite geometry? How should these aliases be classified? How can they be generated and drawn automatically on a computer with the smallest possible number of crossings of their edges on a planar diagram? When will there be aliases in which love is always, or never, reciprocated? And so on.

5. Finite geometries.

We shall write $\mathfrak{F}_u$ for a finite geometry which satisfies axioms (i), (ii), and (iii) of §4 with $n = u^2 + u + 1$ according to (4.4). It is a famous unsolved problem to determine the values of u for which any $\mathfrak{F}_u$ exists, to specify whether such an $\mathfrak{F}_u$ is unique, and to classify the alternatives for $\mathfrak{F}_u$ when $\mathfrak{F}_u$ is not unique. The following relevant facts and justifications may be found in Hall (1967) and Ryser (1963).

THEOREM 3. (Bruck-Ryser). *If* $u \equiv 1, 2$ (modulo 4) *and if u cannot be expressed as the sum of two squares, then* $\mathfrak{F}_u$ *cannot exist.*

In particular, $\mathfrak{F}_6$ does not exist. It is not known whether $\mathfrak{F}_{10}$ exists.

THEOREM 4. *At least one* $\mathfrak{F}_u$ *exists whenever u is a prime power.*

The simplest way of constructing an $\mathfrak{F}_u$ when $u = p^r$ is to consider triples $(x,y,z) \neq (0,0,0)$ of elements x, y, z from the Galois field GF(p^r). Two triples (x,y,z) and $(\lambda x, \lambda y, \lambda z)$ with $0 \neq \lambda \in \mathrm{GF}(p^r)$ represent the same point. Similarly triples $(\lambda\xi, \lambda\eta, \lambda\zeta) \neq (0,0,0)$ represent lines; and we say that the point (x,y,z) lies on the line (ξ,η,ζ) if and only if $x\xi + y\eta + z\zeta = 0$. It is easy to verify that this does provide a finite Desarguesian geometry $\mathfrak{F}_u$.

On the other hand it is possible to construct non-Desarguesian finite geometries $\mathfrak{F}_u$ for infinitely many values of the form $u = p^r$ by means of near-fields and Veblen-Wedderburn systems. A non-Desarguesian geometry cannot be geometrically equivalent to a Desarguesian geometry; so $\mathfrak{F}_u$ is not unique for certain values of u. Various other constructions, such as Hughes planes, are also known. A necessary and sufficient condition for the existence of $\mathfrak{F}_u$ is that $u-1$ mutually orthogonal Latin squares of order u should exist. Euler conjectured that a pair of orthogonal Latin squares could not exist if $u \equiv 2$ (modulo 4): Tarry (1901) verified the conjecture for $u = 6$, but Bose *et al.* (1959, 1960a, 1960b) demonstrated that Euler's conjecture is false for $u \equiv 10$ (modulo 12). No $\mathfrak{F}_u$ has yet been found for which u is not a prime power.

6. The unordered love problem.

I have only one remark to make about the unordered love problem. If the row totals of **A** are all equal (or equivalently, if the column totals of **A** are all equal), then **A** must satisfy (4.5) and (4.6): see Ryser (1965) for a proof of this. Hence to obtain any additional solutions over and above those of the self-dual problem, we must have at least two distinct row totals and at least two distinct column totals.

7. The ordered love problem.

We have seen that the unordered love problem, which includes the self-dual problem, has a large number of graphically inequivalent solutions. This is in striking contrast to the ordered love problem which seems to have relatively few solutions.

We have to solve (3.1), which we can write as

$$\mathbf{A}^2 = \mathbf{D} \tag{7.1}$$

where $\mathbf{D} = (d_{ij})$ has $d_{ij} = 1$ whenever $i \neq j$. Then

$$\mathbf{DA} = \mathbf{A}^3 = \mathbf{AD}. \tag{7.2}$$

Let r_i and c_j $(i, j = 1, 2, ..., n)$ be the row and column sums of $\mathbf{A}$. The (i, j) element of $\mathbf{DA}$ is $c_j + (d_{ii} - 1)a_{ij}$; and the (i, j) element of $\mathbf{AD}$ is $r_i + (d_{jj} - 1)\, a_{ij}$. So by (7.2)

$$r_i - c_j = (d_{ii} - d_{jj})a_{ij}. \tag{7.3}$$

Putting $i = j$ in (7.3) we deduce

$$r_i = c_i. \tag{7.4}$$

So (7.3) becomes

$$c_i - c_j = (d_{ii} - d_{jj})a_{ij}. \tag{7.5}$$

Hence, if $c_i \neq c_j$ we must have $a_{ij} \neq 0$, which implies $a_{ij} = 1$. We may without loss of generality suppose that we have permuted the rows and the columns of $\mathbf{A}$ by the same permutation according to (3.6), so that the rows of $\mathbf{A}$ with equal row totals are grouped together: suppose that there are n_1 rows with equal row totals c_1, n_2 rows with equal totals c_2, ..., n_k rows with equal totals c_k, where $c_1, c_2, ..., c_k$ are all distinct. Then $\mathbf{A}$ will have the partitioned form

$$\mathbf{A} = \begin{bmatrix} ? & \mathbf{j} & \mathbf{j} & \cdots & \mathbf{j} \\ \mathbf{j} & ? & \mathbf{j} & \cdots & \mathbf{j} \\ \cdot & \cdot & \cdot & \cdots & \cdot \\ \mathbf{j} & \mathbf{j} & \mathbf{j} & \cdots & ? \end{bmatrix} \begin{matrix} n_1 \text{ rows} \\ n_2 \text{ rows} \\ \\ n_k \text{ rows} \end{matrix} \tag{7.6}$$

in which $\mathbf{j}$ denotes some rectangular submatrix whose elements are all 1, and **?** denotes an unspecified square submatrix.

We shall show first that $k \leqslant 2$. For suppose that $k \geqslant 3$. We cannot have $n_1 > 1$, for otherwise $1 = d_{12} \geqslant n_2 + n_3 + ... + n_k \geqslant 2$. So $n_1 = 1$. Similarly $n_2 = n_3 = \cdots = 1$. But then each of the submatrices marked **?** contains only a single element 0 or 1; and so $c_j = n$ or $n - 1$. This contradicts the fact that there are at least three distinct values for the c_j. This leaves us with two cases only: $k = 1$ or 2.

Consider the case $k = 2$. If $n_1 > 1$, then $1 = d_{12} \geqslant n_2$; and therefore $n_2 = 1$ and

$n_1 = n-1$. Alternatively, if $n_2 > 1$, we get $n_1 = 1$, $n_2 = n-1$. Without loss of generality we can assume that $n_1 = 1$ and $n_2 = n-1$; and that (7.6) becomes in partitioned form

$$\mathbf{A} = \begin{bmatrix} a_{11} & \mathbf{1}^{\mathrm{T}} \\ \mathbf{1} & \mathbf{A}_{22} \end{bmatrix}, \tag{7.7}$$

where $\mathbf{1}$ is a column of 1's. Now, for $j > 1$, consider $1 = d_{1j} = c_j + a_{11} - 1$. If $a_{11} = 1$, then $c_j = 1$ and $\mathbf{A}_{22} = \mathbf{0}$. This gives the solution (4.2). On the other hand, if $a_{11} = 0$, $c_j = 2$; and so each row and each column of $\mathbf{A}_{22}$ will contain just one 1, that is to say $\mathbf{A}_{22}$ must be a permutation matrix, say $\mathbf{Q}$. Then

$$\mathbf{D} = \mathbf{A}^2 = \begin{bmatrix} \mathbf{1}^{\mathrm{T}}\mathbf{1} & \mathbf{1}^{\mathrm{T}}\mathbf{Q} \\ \mathbf{Q1} & \mathbf{11}^{\mathrm{T}} + \mathbf{Q}^2 \end{bmatrix}; \tag{7.8}$$

whereupon (3.1) implies that $\mathbf{Q}^2$ is a diagonal matrix. Consequently $\mathbf{Q}$ represents a permutation of order 2, which must be a product of disjoint 1-cycles and 2-cycles. Therefore the structure of the community (to within graphical equivalence) is

$$\begin{aligned} &V_1 \rightleftharpoons V_i \quad (i = 2, 3, \ldots, n) \\ &V_{2i} \rightleftharpoons V_{2i+1} \quad (i = 1, 2, \ldots, \tfrac{1}{2}n - \tfrac{1}{2}m - \tfrac{1}{2}) \\ &V_i \rightleftharpoons V_i \quad (i > n-m). \end{aligned} \tag{7.9}$$

Here n and m are of opposite parity; the second line is void if $m = n-1$; and the third line is void if $m = 0$. The friendship theorem is the particular case $m = 0$. These solutions are rather trivial; and hereafter we deal only with the case $k = 1$.

When $k = 1$, we get

$$\mathbf{JA} = \mathbf{AJ} = c\mathbf{J}, \tag{7.10}$$

where $c > 0$ is the common value of any row total or any column total for $\mathbf{A}$. By (7.1) we get

$$\mathbf{DJ} = \mathbf{A}^2\mathbf{J} = c\mathbf{AJ} = c^2\mathbf{J}. \tag{7.11}$$

But all the elements of the ith row of $\mathbf{DJ}$ are $d_{ii} + n - 1$. Hence $d_{ii} = c^2 + 1 - n$, which gives

$$\mathbf{D} = \mathbf{A}^2 = (c^2 - n)\mathbf{I} + \mathbf{J}. \tag{7.12}$$

Hence

$$(\mathbf{D} - c^2\mathbf{I})(\mathbf{D} + n\mathbf{I} - c^2\mathbf{I}) = (\mathbf{D} - c^2\mathbf{I})\mathbf{J} = \mathbf{0}, \tag{7.13}$$

by (7.11) and (7.12). The left-hand side of (7.13) must be the minimal polynomial of $\mathbf{D}$, because $\mathbf{D}$ is not diagonal. So the eigenvalues of $\mathbf{D}$ must be c^2 (with multiplicity α, say) and $c^2 - n$ (with multiplicity $n - \alpha$). However the sum of the eigenvalues of $\mathbf{D}$ equals the trace of $\mathbf{D}$. Thus

$$c^2\alpha + (c^2 - n)(n - \alpha) = (c^2 + 1 - n)n, \tag{7.14}$$

which implies $\alpha = 1$.

Next consider the minimal polynomial for **A**. It cannot be of the first degree, because **D** would be diagonal if **A** were diagonal. Suppose, for the sake of a contradiction, that it is of the second degree, say

$$\mathbf{A}^2 + a\mathbf{A} + b\mathbf{I} = \mathbf{0}. \tag{7.15}$$

Then, by (7.12) and (7.15),

$$a\mathbf{A} = (n - c^2 - b)\mathbf{I} - \mathbf{J}. \tag{7.16}$$

Consideration of the off-diagonal elements in (7.16) shows that $a = -1$ and $a_{ij} = 1$ for all $i \neq j$. But then $1 = d_{1n} \geqslant n - 1$ implies $n = 2$; whence we easily find that

$$\mathbf{A} = \begin{bmatrix} 1 & 1 \\ 1 & 0 \end{bmatrix} \text{ or } \begin{bmatrix} 0 & 1 \\ 1 & 1 \end{bmatrix}, \tag{7.17}$$

contradicting the assumption that the row sums of **A** are all equal. Hence the degree of the minimal polynomial is at least 3. On the other hand, from (7.10) and (7.11),

$$(\mathbf{A} - c\mathbf{I})(\mathbf{A}^2 + n\mathbf{I} - c^2\mathbf{I}) = (\mathbf{A} - c\mathbf{I})\mathbf{J} = \mathbf{0}. \tag{7.18}$$

Therefore the left-hand side of (7.18) is the minimal polynomial of **A**. The eigenvalues of **A** are accordingly c and $\pm\sqrt{(c^2 - n)}$. The multiplicity of the eigenvalue c cannot exceed $\alpha = 1$ (the multiplicity of the eigenvalue $c^2$of $\mathbf{A}^2$). So c is a simple eigenvalue. Let β denote the multiplicity of the eigenvalue $+\sqrt{(c^2 - n)}$; so $n - 1 - \beta$ is the multiplicity of the eigenvalue $-\sqrt{(c^2 - n)}$. We now have, for the narcissism m,

$$m = \text{trace } \mathbf{A} = c + (2\beta + 1 - n)\sqrt{(c^2 - n)}. \tag{7.19}$$

Since $a_{ij} = 0$ or 1 and $d_{ii} = c^2 + 1 - n$, we have

$$0 \leqslant d_{ii} = c^2 + 1 - n = \textstyle\sum_j a_{ij} a_{ji} \leqslant \sum_j a_{ij} = c. \tag{7.20}$$

If $d_{ii} = 0$, then $c^2 - n = -1$; and equating the real parts of (7.19) we get $m = c > 0$; so $a_{ii} = 1$ for some i and $d_{ii} \geqslant 1$, which is a contradiction. Hence we can replace 0 by 1 on the left of (7.20) and conclude that

$$(c^2 - 1)^2 < c^2 - c + 1 \leqslant n \leqslant c^2. \tag{7.21}$$

But c is an integer. So we conclude

$$c = \text{the least integer not less than } \sqrt{n}. \tag{7.22}$$

Now consider (7.19). We have $m = c$ if $n = c^2$ or if $2\beta = n - 1$, the latter being possible only if n is odd. Suppose that neither of these possibilities occurs; so that $m \neq c$. Then $\sqrt{(c^2 - n)}$ must be an integer, say v; and

$$n = c^2 - v^2, \tag{7.23}$$

where

$$0 < v \leqslant \sqrt{(c^2 - c + 1)} \tag{7.24}$$

by (7.21). Further, from (7.19),

$$m \equiv c+(n-1)v^2 \equiv c+(n-1)(c^2-n) \equiv nc \quad \text{(modulo 2)}, \tag{7.25}$$

and

$$m \equiv c \quad \text{(modulo } v). \tag{7.26}$$

Two distinct narcissistic individuals V_i and V_j cannot be connected by an edge; for, if there were an edge $V_i \rightarrow V_j$ then there would be two solutions $k = i$ and $k = j$ to $V_i \rightarrow V_k \rightarrow V_j$. There are c edges leaving V_i and c edges entering V_i, of which c^2+1-n are bibs. One of these bibs connects V_i to itself if V_i is narcissistic. Hence V_i is connected to $2c-c^2+n-2$ non-narcissistic vertices, of which there are $n-m$. Hence

$$m \leqslant c^2-2c+2. \tag{7.27}$$

We assemble the foregoing results as a theorem:

THEOREM 5. *Any solution of* $\mathfrak{L}_n^1$ *which is not graphically equivalent to* (4.2) *or* (7.9) *has the following properties in which c is the least integer not less than* $\sqrt{n}$.

$$n = c^2-v^2 \geqslant c^2-c+1, \tag{7.28}$$

for some integer $v \geqslant 0$. *There are exactly c edges from each vertex and c edges to each vertex; of these c edges,* v^2+1 *are bibs and the remainder are arcs. The narcissism m satisfies*

$$0 \leqslant m \leqslant c^2-2c+2. \tag{7.29}$$

If $v = 0$, *then*

$$m = c. \tag{7.30}$$

If $v > 0$, *then either* (7.30) *holds or else*

$$m \equiv cn \quad \text{(modulo 2)} \quad \textit{and} \quad m \equiv c \quad \text{(modulo } v). \tag{7.31}$$

Let us examine the contrasts between Theorems 2 and 5, the former covering solutions of $\mathfrak{L}_n^4$ that are not geometrically equivalent to (4.3), and the latter covering solutions of $\mathfrak{L}_n^1$ that are not graphically equivalent to (4.2) or (7.9). For Theorem 2, each row and column of **A** has sum $u+1 = c$ and

$$n = u^2+u+1 = c^2-c+1. \tag{7.32}$$

Thus in Theorem 2, the possible values of n are more restricted than in (7.28) of Theorem 5. On the other hand, in Theorem 2 the narcissism may have any value in $0 \leqslant m \leqslant n$, whereas in Theorem 5 (7.30) and (7.31) limit the possible values of m to less than half of the values in $0 \leqslant m \leqslant n$. It seems rather likely that (7.29) could be sharpened considerably: instead of m being of the order of c^2 one might guess that m should always be of the order of c. But we shall see in §8 that (7.30) is not always true, and that m can have more than one value for given n. We now turn to some specific solutions for $\mathfrak{L}_n^1$.

8. Circulants and difference sets.

In this section we shall number the rows and columns of a matrix from 0 to $n-1$ instead of 1 to n. An $n \times n$ matrix **A** is called a *g-circulant* if each row of **A** is obtained by cyclically shifting the elements of the preceding row g places to the right. Thus if the elements of the zeroth row of **A** are $f_0, f_1, \ldots, f_{n-1}$, then

$$a_{ij} = f_{j-gi}, \tag{8.1}$$

where the suffices are treated modulo n. With any circulant we associate its Hall polynomial

$$f(x) = f_0 + f_1 x + \ldots + f_{n-1} x^{n-1}, \tag{8.2}$$

in which x is an indeterminate satisfying

$$x^n = 1. \tag{8.3}$$

Suppose that **B** is an h-circulant with a Hall polynomial

$$\varphi(x) = \varphi_0 + \varphi_1 x + \ldots + \varphi_{n-1} x^{n-1}. \tag{8.4}$$

The (i, j)-element in the product $\mathbf{C} = \mathbf{AB}$ is

$$c_{ij} = \textstyle\sum_k a_{ik} b_{kj} = \sum_k f_{k-gi} \varphi_{j-kh} = \text{coefficient of } x^{j-ghi} \text{ in } f(x^h)\varphi(x). \tag{8.5}$$

Hence **AB** is a gh-circulant with Hall polynomial $f(x^h)\varphi(x)$. In particular, $\mathbf{A}^2$ is a g^2-circulant with Hall polynomial $f(x^g)f(x)$.

From (7.12) we see that $\mathbf{A}^2$ is a 1-circulant if $c^2 > n$, while $\mathbf{A}^2$ is a g-circulant for all values of g if $c^2 = n$. We can obtain circulant solutions of $\mathfrak{L}_n^1$ as follows. If $n = c^2$, we may take

$$f(x) = 1 + x + \ldots + x^{c-1}. \tag{8.6}$$

Then $f(x^c)f(x) = 1 + x + \ldots + x^{n-1}$; so **A**, the c-circulant with Hall polynomial (8.6) satisfies (7.12). In this case $m = c$, as already noted. On the other hand, if $c^2 > n$, we require **A** to be a g-circulant with $g^2 \equiv 1$ (modulo n) and

$$f(x^g)f(x) = (c^2 - n + 1) + x + x^2 + \ldots + x^n. \tag{8.7}$$

Again we can satisfy (8.7) in the case $n = c^2 - 1$ by taking **A** to be the c-circulant with Hall polynomial (8.6). This time we find $m = 2c - 2$.

Difference sets provide circulant solutions when $n = c^2 - c + 1$. A set of residues modulo n is called a *difference set* if every residue, except 0, can be expressed uniquely as the difference of two members of the set. Since there are $n - 1$ residues apart from zero, and $c(c-1)$ ordered pairs arising from a difference set of c members, we have $n - 1 = c(c-1)$. We write

$$f(x) = \textstyle\sum \varepsilon_i x^i, \tag{8.8}$$

where $\varepsilon_i = 1$ or 0 according as the residue i does or does not belong to the difference

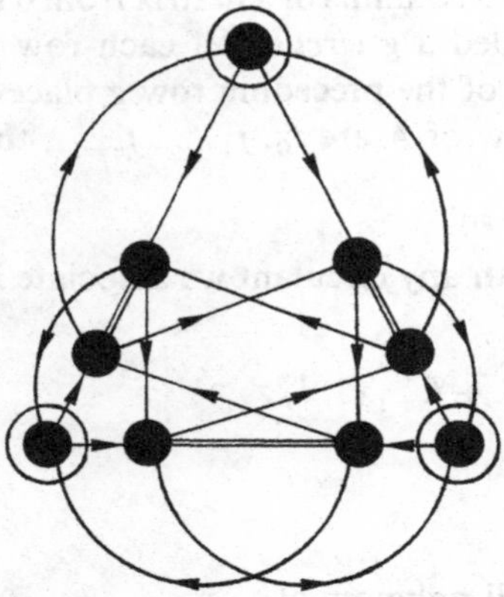

Figure 17. $\mathfrak{G}_9^1, m = 3$.

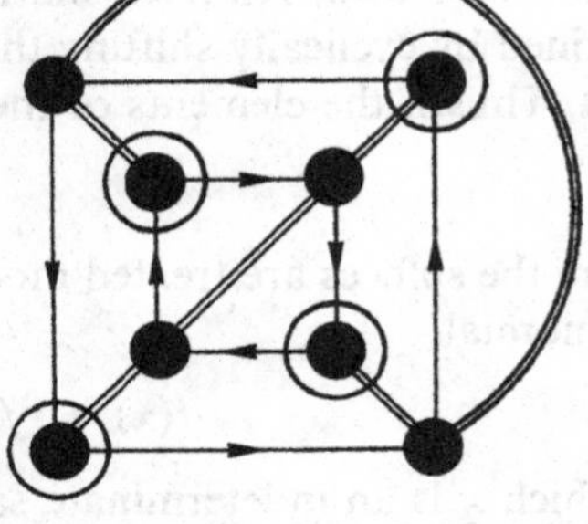

Figure 18. $\mathfrak{G}_8^1, m = 4$.

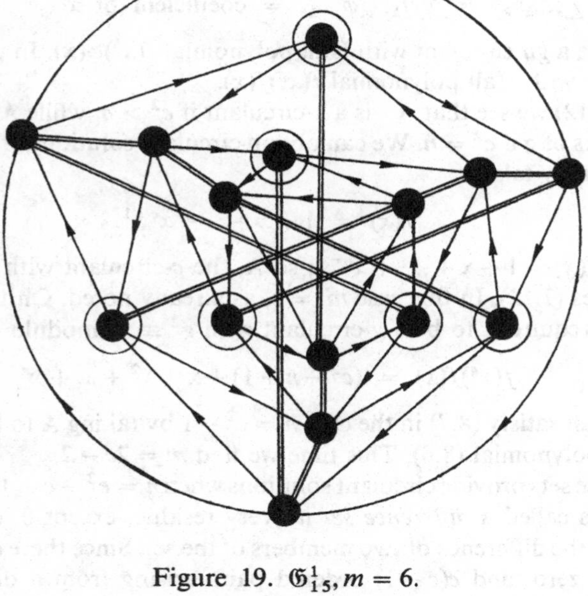

Figure 19. $\mathfrak{G}_{15}^1, m = 6$.

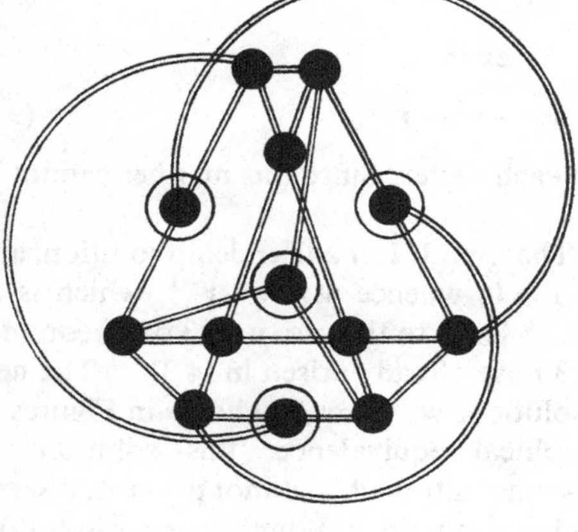

Figure 20. $\mathfrak{G}^1_{13} = \mathfrak{G}^4_{13}, m = 4.$

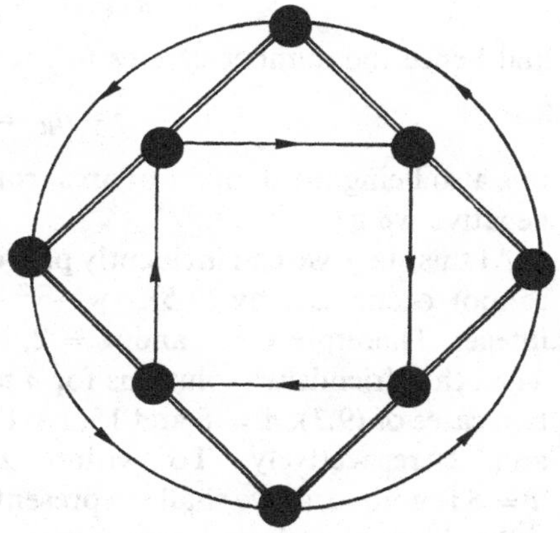

Figure 21. $\mathfrak{G}^1_8, m = 0.$

set. Then clearly

$$f(x^{-1})f(x) = c + x + x^2 + \dots + x^n; \tag{8.9}$$

and $f(x)$ is the Hall polynomial of a g-circulant with $g = -1$. This gives $m = c$ and $\beta = \frac{1}{2}(n-1)$. Solutions arising from difference sets, when they exist, always yield symmetrical matrices **A**, because $g = -1$. Hence these solutions of $\mathfrak{L}^1_n$ are simultaneously solutions of $\mathfrak{L}^4_n$.

Figure 17 shows the circulant solution (8.6) for $n = 9$; Figures 18 and 19 the circulant solutions (8.7) for $n = 8$ and $n = 15$; and Figures 16 and 20 the solutions (8.8) arising from the difference sets $\{0,1,3\}$ when $n = 7$ and $\{0,1,4,6\}$ when $n = 13$.

9. The ordered love problem with $m = 0$.

When $m = 0$, we see from (7.19) that (7.23) must hold, because $c > 0$ and $m \neq c$. Hence

$$\beta = \tfrac{1}{2}(n - 1 - c/v), \tag{9.1}$$

which is a non-negative integer. So $c = vw$ for some integer w, and (7.23) yields

$$n = v^2(w^2 - 1). \tag{9.2}$$

Moreover, by (9.1), w and n must be of opposite parity; and therefore, by (9.2), v and w cannot both be even. We shall also see presently that $w > v$. The only possible values of n are therefore

$$n = 3, 8, 24, 32, 35, 48, 63, 80, 96, 99, \dots \tag{9.3}$$

The number of bibs attached to each vertex is

$$d_{ii} \;=\; c^2+1-n \;=\; v^2+1; \tag{9.4}$$

and hence the number of arcs to each vertex is

$$c-d_{ii} \;=\; vw-v^2-1, \tag{9.5}$$

this also being the number of arcs from each vertex. Since this number cannot be negative we get $w > v$.

At this stage we can incidently prove Theorem 1. For a friendship solution, arcs cannot occur. So, by (9.5), $vw-v^2-1=0$; whence $w=v+v^{-1}$, which is an integer. Therefore $v = 1$ and $w = 2$, which leads to the case $n = 3$ of Theorem 1. The other friendship solutions for $n > 3$ have already arisen in (4.3). The next two cases of (9.3), $n = 8$ and 15, have solutions with arcs, as shown in Figures 21 and 2 respectively. To within graphical equivalence, this solution for $n = 8$ is unique, as we shall see presently; and, although I cannot prove it, it seems rather likely that the solution for $n = 15$ is also unique. Whether or not solutions for (9.3) with $n = 24, 32, \ldots$ exist and are unique I cannot say.

To investigate these questions further we make use of a lemma:

LEMMA. *Any quadrilateral in the graph of a solution of* $\mathfrak{L}_n^1$ *must have at least one pair of opposite sides which are arcs with opposite directions.*

This is easily proved; because if the condition is not satisfied, there will be a pair of diagonally opposite vertices V_i, V_j of the quadrilateral such that V_j can be reached from V_i in two distinct ways along the sides of the quadrilateral in contradiction of $S_1(i, j)$.

The solutions $n = 8$ and 15 of (9.3) both have $v = 1$; so, by (9.4), there are exactly 2 bibs attached to each vertex. It follows from Euler's theorem, that the *bibgraphs* (i.e. the graphs obtained by deleting all arcs) consist of a set of disjoint cycles.

Consider now the case $n = 8$. The bibgraph cannot contain a 4-cycle, for that would violate the Lemma. Suppose for the sake of a contradiction that the bibgraph contains a 3-cycle, through the vertices V_1, V_2, V_3 say. According to (9.5), each of these three vertices has $w-2=1$ arc to it and 1 arc from it. The other ends of these arcs must be at distinct vertices, for otherwise the Lemma would be violated. So there are 6 more vertices besides V_1, V_2, V_3. Thus the graph has at least 9 vertices in contradiction of $n = 8$. This proves that a 3-cycle is impossible in the bibgraph. The disjoint cycles of the bibgraph could only be of lengths (3,5) or (4,4) or (8); and we have just dismissed 3 and 4 as possible lengths. So the bibgraph must be an 8-cycle. Label the vertices V_1, V_2, ..., V_8 consecutively around this 8-cycle treating the suffices modulo 8. The arc from V_i cannot go to V_{i+1} or V_{i-1} because it would then coincide with a bib, nor to V_{i+3} or V_{i-3} because that would violate the Lemma. Suppose, for the sake of a contradiction, that we have $V_i \to V_{i+4}$. Then $V_{i+4} \to V_i$ is impossible because that would create a bib between

V_i and V_{i+4}; $V_{i+4} \to V_{i+6}$ and $V_{i+4} \to V_{i+2}$ would violate the Lemma; and all other arcs from V_{i+4} are already forbidden by the previous sentence. So V_{i+4} can have no arc from it; and this contradiction means that we must disallow the assumption $V_i \to V_{i+4}$. We conclude that all arcs must be of the form $V_i \to V_{i+2}$ or $V_i \to V_{i-2}$. By reversing, if necessary, the order of naming the vertices around the bibgraph, we may assume that to within graphical equivalence $V_1 \to V_3$. This implies $V_1 \to V_3 \to V_5 \to V_7 \to V_1$ to avoid extra bibs. Thereupon $V_2 \to V_4$ is forbidden by the Lemma. Hence $V_2 \to V_8 \to V_6 \to V_4 \to V_2$. This completes the proof that the solution in Figure 21 is unique to within graphical equivalence.

Next consider the case $n = 15$; and suppose, for the sake of a contradiction that the bibgraph contains a 3-cycle $V_1 \rightleftharpoons V_2 \rightleftharpoons V_3 \rightleftharpoons V_1$. By (9.5), each vertex has 2 arcs from it and 2 arcs to it. For the same reasons as in the case $n = 8$, the arcs to and from the 3-cycle lead to 12 further distinct vertices; and we may number the vertices so that

$$\begin{array}{llll} V_1 \to V_4, & V_1 \to V_5, & V_1 \leftarrow V_6, & V_1 \leftarrow V_7 \\ V_2 \to V_8, & V_2 \to V_9, & V_2 \leftarrow V_{10}, & V_2 \leftarrow V_{11} \\ V_3 \to V_{12}, & V_3 \to V_{13}, & V_3 \leftarrow V_{14}, & V_3 \leftarrow V_{15}. \end{array} \tag{9.6}$$

Now consider the unique solution $k(1, 4)$ of $V_1 \to V_k \to V_4$. We must have $k \leqslant 5$ because the only edges leading from V_1 lead to V_2, V_3, V_4, V_5. Further $k = 2, 3$ are impossible because no edges from V_2 or V_3 terminate at V_4. Also $k = 4$ is impossible because $m = 0$. Hence $k = 5$. So we have

$$V_1 \to V_5 \to V_4. \tag{9.7}$$

Similarly we find

$$V_1 \to V_4 \to V_5. \tag{9.8}$$

Hence

$$V_4 \rightleftharpoons V_5 \tag{9.9}$$

is a bib. Likewise, by considering the unique solutions k of $V_6 \to V_k \to V_1$ and $V_7 \to V_k \to V_1$, we find that

$$V_6 \rightleftharpoons V_7 \tag{9.10}$$

must be a bib. Next consider the unique solution of $V_4 \to V_k \to V_1$. We find that $k = 6$ or 7; and, if necessary renumbering the vertices, we may assume

$$V_4 \to V_6. \tag{9.11}$$

Similarly the solution of $V_5 \to V_k \to V_1$ can only be $k = 6$ or 7. If $k = 7$, then by (9.9), (9.10), (9.11) the quadrilateral V_4, V_5, V_6, V_7 violates the Lemma. Hence

$$V_5 \to V_6. \tag{9.12}$$

But now $V_1 \to V_k \to V_6$ has two distinct solutions $k = 4$ and $k = 5$. This contradiction shows that the bibgraph does not contain a 3-cycle.

Since the bibgraph cannot contain a 4-cycle, the bibgraph can only be the union of disjoint cycles of lengths

$$(5,5,5) \text{ or } (5,10) \text{ or } (6,9) \text{ or } (7,8) \text{ or } (15). \tag{9.13}$$

Suppose that the bibgraph contains a 5-cycle $V_1 \rightleftharpoons V_2 \rightleftharpoons V_3 \rightleftharpoons V_4 \rightleftharpoons V_5 \rightleftharpoons V_1$. There cannot be an arc of the form $V_1 \to V_3$, for otherwise the quadrilateral V_1, V_3, V_4, V_5 would violate the Lemma. Similarly $V_1 \to V_4$ leads to an inadmissible quadrilateral V_1, V_2, V_3, V_4. We already have bibs $V_1 \rightleftharpoons V_2$ and $V_1 \rightleftharpoons V_5$; and therefore there cannot be a further arc of the form $V_1 \to V_2$ or $V_1 \to V_5$. Hence all the 10 arcs that leave the vertices of the 5-cycle terminate in the residual set $\{V_6, V_7, ..., V_{15}\}$. Similarly all 10 arcs that terminate in the 5-cycle originate from the residual set. In particular, let W_i $(i = 1,2,...,5)$ be the unique solution $k = k(i, i+1)$ of

$$V_i \to V_k = W_i \to V_{i+1}; \tag{9.14}$$

so that $W_i \in \{V_6, V_7, ..., V_{15}\}$. We claim that $W_1, W_2, ..., W_5$ are all distinct vertices; for the following assumptions imply the following quadrilaterals violating the Lemma

Assumption	Violating quadrilateral
$W_1 = W_2$	$V_1, V_2, V_3, W_1 = W_2$
$W_1 = W_3$	$V_1, V_4, V_5, W_1 = W_3$
$W_1 = W_4$	$V_1, V_2, V_5, W_1 = W_4$

while the assumption $W_1 = W_5$ implies that $V_1 \rightleftharpoons W_1 = W_5$ and hence that 3 bibs are attached to V_1. The assertion that all the W_i are distinct now follows by symmetry. Likewise, if X_i $(i = 1,2,...,5)$ are the unique solutions $k = k(i+1, i)$ of

$$V_{i+1} \to V_k = X_i \to V_i, \tag{9.15}$$

then $X_1, X_2, ..., X_5$ are all distinct vertices, because the 5-cycle of bibs can be described in either direction. Next we claim that none of the W_i can be the same as any of the X_j, for the following assumptions lead to the following violating quadrilaterals:

Assumption	Violating quadrilateral
$W_1 = X_2$	$V_1, V_2, V_3, W_1 = X_2$
$W_1 = X_3$	$V_1, V_2, V_3, W_1 = X_3$
$W_1 = X_4$	$V_1, V_2, V_3, W_1 = X_4$
$W_1 = X_5$	$V_1, V_2, V_5, W_1 = X_5$

while $W_1 = X_1$ implies that $V_1 \rightleftharpoons W_1 = X_1$ is a third bib attached to V_1. By cyclic symmetry, no W_i can coincide with any X_j. It follows that the W_i and the X_i exhaust the set $\{V_6, V_7, ..., V_{15}\}$; so by renaming the vertices W_i, X_i we can assume without loss of generality that

$$\begin{aligned} &V_1 \to V_6 \to V_2 \to V_7 \to V_1 \\ &V_2 \to V_8 \to V_3 \to V_9 \to V_2 \\ &V_3 \to V_{10} \to V_4 \to V_{11} \to V_3 \\ &V_4 \to V_{12} \to V_5 \to V_{13} \to V_4 \\ &V_5 \to V_{14} \to V_1 \to V_{15} \to V_5. \end{aligned} \tag{9.16}$$

The vertices $V_1, V_2, \ldots, V_5$ now have all their edges attached; so all remaining edges are between members of the set $\{V_6, V_7, \ldots, V_{15}\}$, and to each of these vertices we still have to attach 2 bibs and 1 incoming arc and 1 outgoing arc.

Now consider the solution $k(1, 6)$ of $V_1 \to V_k \to V_6$. We cannot have $1 < k \leqslant 5$ because we have exhausted all outgoing arcs from $V_2, V_3, \ldots, V_5$; and $k = 1$ is forbidden because $m = 0$. The only remaining outgoing arc from V_1 is $V_1 \to V_{15}$. Hence $k(1, 6) = 15$ and $V_{15} \to V_6$. A similar argument shows that $k(1, 15) = 6$ and $V_6 \to V_{15}$. Thus $V_6 \rightleftharpoons V_{15}$ is a bib. Again $k(9, 2) \neq 2$ because $m = 0$, and then $k(9, 2) > 5$ in view of the incoming edges to V_2 in (9.16). Hence $k(9, 2) = 6$. Similarly $k(6, 2) = 9$; and therefore $V_6 \rightleftharpoons V_9$ is a bib. Therefore, by cyclic symmetry, we now have a 10-cycle in the bibgraph, namely

$$V_6 \rightleftharpoons V_9 \rightleftharpoons V_{10} \rightleftharpoons V_{13} \rightleftharpoons V_{14} \rightleftharpoons V_7 \rightleftharpoons V_8 \rightleftharpoons V_{11} \rightleftharpoons V_{12} \rightleftharpoons V_{15} \rightleftharpoons V_6. \tag{9.17}$$

This prohibits the first possibility (5,5,5) in (9.13).

The edges that have now been allocated enable us to write down the solutions $k(1, j)$ as follows:

$j =$	2	3	4	5	6	7	8	9	10	11	12	13	14	15
$k(1,j) =$	6	2	5	15	15	2	2	6	?	?	15	5	5	6

(9.18)

The solutions $k(1, j)$ for the values $j = 10, 11$ must belong to the set $\{2, 5, 6, 15\}$ since these provide the only vertices directly accessible from V_1. Each member of this set can appear at most 4 times, since $c = 4$; but it can only appear 3 times if it arises as a bib from V_1, the inadmissible fourth occurence corresponding to a return to V_1. Thus 2 and 5, arising from bibs, already appear 3 times in (9.18). Hence the missing entries in (9.18) (represented by ?) must be 6 and 15 each taken once. But $k(1, 11) = 6$ implies $V_6 \to V_{11}$ which gives the violating quadrilateral V_2, V_6, V_8, V_{11}. Hence $k(1, 11) = 15$ and $k(1, 10) = 6$; and therefore $V_6 \to V_{10}$ and $V_{15} \to V_{11}$. By cyclic symmetry, we now have all the remaining arcs

$$\begin{aligned} &V_6 \to V_{10},\ V_{15} \to V_{11} \\ &V_8 \to V_{12},\ V_7 \to V_{13} \\ &V_{10} \to V_{14},\ V_9 \to V_{15} \\ &V_{12} \to V_6,\ V_{11} \to V_7 \\ &V_{14} \to V_8,\ V_{13} \to V_9. \end{aligned} \tag{9.19}$$

This completely specifies the graph; so we can write down its incidence matrix **A** and check that we do indeed have a solution of $\mathfrak{L}^1_{15}$ with $m = 0$. This is the graph illustrated in Figure 2.

Analysis of the last three cases (6,9), (7,8), and (15) in (9.13) is more difficult. I have only some incomplete results that are not worth reporting here, though they incline me to the belief that none of these cases can yield a solution of $\mathfrak{L}^1_{15}$. If that conjecture is correct, then Figure 2 is the unique asymmetric solution of $\mathfrak{L}^1_{15}$ with $m = 0$.

10. Typographical acknowledgment: OUCS Lasercomp.

This article has been typeset on a Lasercomp at Oxford University Computing Service. I am indebted to Gillian Holiday, who drew the diagrams, and to Giuseppe Mazzarino, who helped with the typesetting.

References.

Birkhoff, G. (1946) Three observations on linear algebra. *Rev. Univ. Tucumán* A **5**, 147–151.

Bose, R.C. & Shrikande, S.S. (1959) On the falsity of Euler's conjecture about the non-existence of two orthogonal Latin squares of order $4t+2$. *Proc. Nat. Acad. Sci. U.S.A.* **45**, 734–737.

Bose, R.C. & Shrikande, S.S. (1960) On the construction of sets of mutually orthogonal Latin squares and the falsity of a conjecture of Euler. *Trans. Amer. Math. Soc.* **95**, 191–209.

Bose, R.C., Shrikande, S.S. & Parker, E.T. (1960) Further results on the construction of mutually orthogonal Latin squares and the falsity of Euler's conjecture. *Canad. J. Math.* **12**, 189–203.

Carnegie, D.B. (1953) *How to win friends and influence people.* Kingswood, Surrey.

Hall, M. (1959) *The theory of groups.* Macmillan.

Hall, M. (1967) *Combinatorial theory.* Blaisdell Publishing Company.

Horn, A. (1954) Doubly stochastic matrices and the diagonal of a rotation matrix. *Amer. J. Math.* **76**, 620–630.

Lam, C.W.H. (1977) On some solutions of $\mathbf{A}^k = d\mathbf{I} + \lambda\mathbf{J}$. *J. Comb. Th.* (A) **23**,140–147.

Liu, C.L. (1968) *Introduction to combinatorial mathematics.* McGraw Hill.

Ryser, H.J. (1963) Combinatorial mathematics. *Carus Mathematical Monographs* **14**. Mathematical Association of America.

Tarry,G. (1901) Le problème de 36 officieurs. *Compt. Rend. Assoc. Franc. Avance. Sci. Nat.* **2**,170–203.

Wilf, H.S. (1971) The friendship theorem. *Combinatorial Mathematics and its applications* (ed. D.J.A. Welsh) 307–309. Academic Press.

MAXIMUM SETS IN FINITE PROJECTIVE SPACES

J.W.P. Hirschfeld
Mathematics Division, University of Sussex, Falmer,
Brighton BN1 9QH, U.K.

Let PG(d,q) be projective space of d dimensions over the finite field of q elements. Let

$$\theta_d = |PG(d,q)| = q^d + q^{d-1} + \ldots + q+1 = (q^{d+1}-1)/(q-1).$$

A subspace of dimension ℓ is denoted π_ℓ. A π_0 is a *point*, a π_1 is a *line*, a π_2 is a *plane*, a π_3 is a *solid*, a π_{d-1} in PG(d,q) is a *prime* or *hyperplane*. A point with coordinate vector X is denoted P(X). In particular, for $0 \le i \le d$,

$U_i = P(0,\ldots,0,1,0,\ldots,0)$ where 1 is in the (i+1)-th place;
$U = P(1,1,\ldots,1)$.

C = (A,B) means that C is the highest common factor of A and B. For basic properties of PG(d,q), see Hirschfeld 1979.

A (k; r,s; d,q)-*set* K satisfies the following properties:

(a) $K \subset PG(d,q)$ with $|K| = k$;

(b) there is a π_s containing r points of K, but no π_s contains r + 1 points of K;

(c) there is a π_{s+1} containing r+2 points of K.

The essential defining property is (b). Property (c) only comes into play in some aspects of (k;r)-sets. We may think of (b) as maximising r and (c) as maximising s.

To make the definition more immediate, we look at the examples that will be considered in this survey.

1) A (k; 2,1; d,q)-set K is a set of k points with at most two on a line. When d=2, the set K is a (*plane*) k-*arc* and, when d>2, the set K is a k-*cap*.

2) A (k; n, 1; d,q)-set K is a set of k points with at most n on a line. When d=2, the set K is a (*plane*) (k;n)-*arc*;

when $d>2$, K is a $(k;n)$-*cap*.

3) A $(k;\ n,\ d-1;\ d,\ q)$-set K is a set of k points with at most n in a prime. For $d>2$, K is a $(k;n)$-*arc*.

4) A $(k;\ d,\ d-1;\ d,\ q)$-set K is a set of k points at most d of which lie in a prime. Alternatively, every set of $d+1$ points of K is independent. In terms of the coordinate vectors of the points of K, every $d+1$ points of K form a basis of the corresponding $(d+1)$-dimensional vector space. K is called a k-*arc*.

5) A $(k;\ r,\ r-1;\ d,\ q)$-set K is a set of k points at most r in a π_{r-1} but some $r+2$ in a π_r; that is, every set of $r+1$ points is independent but some set of $r+2$ points is dependent. K is called a $(k;r)$-*set* or k-*set of kind* r.

For a generalization of $(k;\ r,s;\ d,\ q)$-sets from sets of points to sets of spaces π_ℓ, see Hirschfeld 1979, Chapter 3.

For general properties of $PG(n,q)$, see Segre 1960, 1967, Hirschfeld 1979.

Problems.

I. Find $m(r,s;\ d,q)$, the maximum value of k for which a $(k;\ r,\ s;\ d,q)$-set exists.

II. Classify the sets K of this maximum size.

III. Classify the sets K which are maximal with respect to inclusion; that is, the $(k;\ r,\ s;\ d,q)$-sets not contained in a $(k+1;\ r,s;\ d,q)$-set. Such sets K are *complete*.

For brevity we write

$$m(n,1;\ d,q) = m_n(d,q),$$
$$m(r,r-1;\ d,q) = M_r(d,q),$$
$$m(d,d-1;\ d,q) = m(d,q)\ .$$

Hence

$$M_d(d,q) = m(d,q),$$
$$M_2(d,q) = m_2(d,q),$$
$$M_2(2,q) = m_2(2,q) = m(2,q).$$

Other authors use a similar but not identical notation. In particular, many replace either r or d by that number plus one.

The writers on these problems almost all belong to one of two schools, one American and one European. The former is that of Bose, the statistician and combinatorialist; the latter is that of Segre, the geometer. The problems were first considered by Bose for their statistical interest and later for their coding theory interest. Most further developments were given by Segre and his school using geometric methods.

In this paper the bulk of results concern Problem I. Most give an upper bound for $m(r,s;d,q)$. The surprising fact is how few exact values of this constant are known. For example, $m_2(d,q)$ is only known exactly for $(d,q) = (2,q),(3,q),(d,2),(4,3),(5,3)$. The results in §§1-6 are generally most powerful for small d and large q, whereas those of §7 are applicable for large d and small q.

The statistical connection

$M_r(d,q)$ is the maximum number of factors which can be accommodated in a symmetrical factorial design in which each factor is at q levels and the blocks are of size q^{d+1}, so that no $(r+1)$-factor or lower order interaction is confounded.

(Bose 1947)

For a fractionally replicated design $1/q^k \times q^s$, consisting of a single block with q^{d+1} plots, where $s = d + 1 + k$, it is required that no t-factor or lower order interaction should be aliased with a t-factor or lower order interaction; then the maximum value of s is $M_{2t-1}(d,q)$. If it is required that no t-factor or lower order interaction should be aliased with a $(t+1)$-factor or lower order interaction, then the maximum value of s is $M_{2t}(d,q)$.

(Bose 1961, Bose & Srivastava 1964)

The coding theoretical connection

For an (n,k) linear code over a q-ary alphabet, where k is the number of information symbols and $d+1 = n-k$ is the number of check symbols, the maximum value of n for which t errors can be corrected with certainty is $M_{2t-1}(d,q)$. The maximum value of n for which t errors can be corrected with certainty and $t+1$ errors can be detected is $M_{2t}(d,q)$.

(Bose 1961)

For further connections, see Bose and Ray-Chaudhuri 1960a, 1960b, Hill 1978a, MacWilliams and Sloane 1977, Chapter 11, Tallini 1960, 1963.

Now a historical survey of results is given. Some of the papers dated 1983 have not yet appeared. In any section the results for the same constant are in improving order.

CONTENTS

§1 r=2, s=1, d=2: plane k-arcs

$$m(2,q) = q + 1, \quad q \text{ odd}$$
$$m(2,q) = q + 2, \quad q \text{ even}$$

(Bose 1947)

An m(2,q)-arc is an *oval*.

(1) For q odd, an oval is a conic.

(Segre 1954, 1955a)

(2) For q even, a conic plus its nucleus (the meet of its tangents) is an oval.

(Bose 1947)

(3) For q=2,4,8, every oval is as in (2).

(Segre 1957)

(4) For q=16,32, or 2^h with h≥7, there exists an oval K other than that in (2). For q≠16, K may be taken as $\{U_1, U_2\} \cup \{P(1,t,t^{2^n}) \mid t \in GF(q),\ (n,h)=1,\ n\neq 1,\ h-1\}$.

(Segre 1957, Hall 1975)

(5) For q even, an oval has a transitive collineation group if and only if $q=2$, 4 or 16.

(Korchmaros 1978)

A crucial question which affects problems in higher dimensions is the following: for what values of k is a k-arc necessarily contained in an oval?

(6) When q is even and $t^2-2t+[\frac{1}{2}t]<q$, then a k-arc, where $k=q+2-t$, is contained in an oval.

(Segre 1959)

(7) When q is even and $k>q-\sqrt{q}+1$, then a k-arc is contained in an oval.

(Segre 1967)

(8) When q is odd and $16t^2+t-37\leq q$ with $t\geq 1$, then a k-arc, where $k=q+2-t$, is contained in a conic.

(Segre 1959)

(9) When q is odd and $k>q-\frac{1}{4}\sqrt{q}+\frac{7}{4}$, then a k-arc is contained in an oval.

(Segre 1967)

Results (6) - (9) are not best possible and can be improved by calculation for small q.

(10) There exists a function $f(q)$ such that no complete k-arc exists with $q-f(q)<k<m(2,q)$; so such a k-arc is contained in an oval. For small q,

q	2	3	4	5	7	8	9	11	13
$f(q)$	2	3	4	5	1	2	1	1	1

(11) There exists a function $g(q)$ such that any k-arc with $k<g(q)$ is incomplete, but there exists a complete $g(q)$-arc. For small q,

q	2	3	4	5	7	8	9	11	13
$g(q)$	4	4	6	6	6	6	6	7	8

(12) For $q\geq 8$ and even, $g(q)\leq\frac{1}{2}(q+4)$.

(Segre 1967)

(13) For $q\geq 9$ and odd, $g(q)\leq\frac{1}{2}(q+3)$.

(Pellegrino 1981,1983)

For further references, see Hirschfeld 1979.

§2 r=n, s=1, d=2: plane (k;n)-arcs with 2<n<q

$m_n(2,q) \le (n-1)q + n$

(Barlotti 1956)

$m_n(2,q) \le (n-1)q + n-2$ for $(q,n)=1$

(Barlotti 1956)

$m_3(2,9) \le 19$

(Cossu 1961)

$m_n(2,q) = (n-1)q + n$ for $q=2^h$, $n=2^e$

(Denniston 1969)

$m_3(2,q) \le 2q + 1$

(Thas 1975)

$m_n(2,q) \le (n-1)q + n-3$ for $4 \le n < q$, $(q,n) = 1$

(Lunelli and Sce 1964)

$m_n(2,q) \le (n-1)q + n-4$ for $9 \le n < q$, $(q,n) = 1$

(Lunelli and Sce 1964)

$m_n(2,q) \le (n-1)q + q-n$ for $n \le 2q/3$, $(q,n) = 1$

(Hill 1983b)

$m_n(2,q) \le (n-1)q + \frac{1}{2}(q+n - n/q) - \frac{1}{2}[(q+n - n/q)^2 - 4(n^2-n)]^{\frac{1}{2}}$
for $n > 2q/3$, $(q,n) = 1$

(Hill 1983b)

$m_n(2,q) \le (n-1)q + q - \sqrt{q} - 1$ for $n = q-1$

(Hill 1983b)

$m_n(2,q) < (n-1)q + q - 3/2 - (q+5/4)^{\frac{1}{2}}$ for $n = q-1$, $q \ge 8$

(Hill 1983b)

$m_n(2,q) \le (n-1)q + q-3/2 - (2q + 7/4)^{\frac{1}{2}}$ for $n = q-2$

(Hill and Mason 1981)

(a) *Small* n,q

n=3	$m_3(2,q) = 2q+1$	for q = 4,5,7
	$15 \le m_3(2,8) \le 17$	
	$m_3(2,9) = 17$	
n=4	$m_4(2,q) = 3q+1$ for q = 5,7,9	
	$m_4(2,8) = 28$	
n=5	$m_5(2,q) = 4q+1$ for q = 7,8,9	
n=6	$m_6(2,q) = 5q + (6,q)$ for q = 7,8,9	

n=7 $6q+1 \le m_7(2,q) \le 6q+2$ for $q = 8,9$

n=8 $65 \le m_8(2,9) \le 66$

For the source of these values, see Mason 1982.

(b) *Lower bounds*

$(\sqrt{q}+1)^2 \le m_3(2,q)$

(Waterhouse 1969)

$(n-1)q+1 \le m_n(2,q)$ for q square, $n = \sqrt{q}+1$

(Bose 1959)

$(n-1)q+1 \le m_n(2,q)$ for q odd, $n = \frac{1}{2}(q+1)$, $\frac{1}{2}(q+3)$

(Barlotti 1965)

$(n-1)q+1 \le m_n(2,q)$ for $n=q-1$

(Hill and Mason 1981)

$(n-1)q+2 \le m_n(2,q)$ for q even, $n=q-2$

(Hill and Mason 1981)

$(n-1)q + n-\sqrt{q} + \sqrt{q}(n-q) \le m_n(2,q)$ for q square

(Hill and Mason 1981)

$(n-1)q + \sqrt{q} \le m_n(2,q)$ for q square, $n = q-\sqrt{q}$

(Mason 1982)

$(n-1)q + q - 2\sqrt{q} - 1 \le m_n(2,q)$ for q square, $n = q-1$

(Hill 1983)

$(n-1)q + q-n \le m_n(2,q)$ for $q-n \mid q$

(Mason 1982)

The lower bound for $m_3(2,q)$ derives from the maximum number of points on an elliptic cubic curve: see Hirschfeld 1983b. For further results see Hirschfeld 1979, Wilson 1974, 1982. The last result is from a preprint.

§3 <u>r=2, s=1, d>2: k-caps</u>

$m_2(3,q) = q^2+1$ for q odd

(Bose 1947)

$m_2(3,q) = q^2+1$ for $q=4$

(Seiden 1950)

$m_2(3,q) = q^2 + 1$ for q even, $q>2$

(Qvist 1952)

$m_2(d,2) = 2^d$

(Bose 1947)

$m_2(4,3) = 20$

(Pellegrino 1970)

$m_2(5,3) = 56$

(Hill 1973)

An $m_2(d,q)$-cap is an *ovaloid.*

(1) For q odd or $q=4$, an ovaloid in $PG(3,q)$ is an elliptic quadric.

(Barlotti 1955, Panella 1955)

(2) For $q=2^{2e+1}$, $e \geq 1$, there exists an ovaloid in $PG(3,q)$ which is not an elliptic quadric.

(Tits 1962)

(3) In $PG(d,2)$, an ovaloid is the complement of a prime.

(Segre 1959)

(4) There are nine projectively distinct ovaloids in $PG(4,3)$.

(Hill 1983a)

(5) The ovaloid in $PG(5,3)$ is projectively unique.

(Hill 1978a)

(6) There exists a function $h(q)$ such that in $PG(3,q)$, $q>2$, a k-cap with $k \geq q^2-h(q)$ is contained in an ovaloid and there exists a complete k'-cap with $k' = q^2-h(q) - 1$. Then

(a) for q odd, $h(q) \geq q-19$;

(Segre 1959)

(b) for $q \geq 7$ and odd, $h(q) \geq q-7$;

(Barlotti 1965)

(c) for $q=3$ or 5, $h(q) \geq 0$;

(Barlotti 1965)

(d) for q even, $h(q) > \frac{1}{2}\sqrt{q}-1$;

(Segre 1967)

(e) for q odd, $h(q) \geq cq^{3/2}$, where c is some constant $< 1/4$ and $q > q_0(c)$ (Segre 1967)

(f) for $q \geq 67$ and odd, $h(q) > \frac{1}{4}q^{3/2}-2q$.

(Hirschfeld 1983a)

(a) *Inequalities*

$q^2 m_2(d-3,q) + 1 \leq m_2(d,q)$ for $d \geq 4$

(Segre 1959)

$m_2(d,q) \leq q m_2(d-1,q) - (q+1)$ for $d \geq 4$, $q>2$

(Hill 1978a)

(b) *Upper bounds*

(i) d=4, q odd

$m_2(4,q) \le q^3$

(Tallini 1956)

$m_2(4,q) \le q^3-1$ for $q=5$

(Barlotti 1957)

$m_2(4,q) \le q^3-q^2+20q-50$

(Segre 1959)

$m_2(4,q) \le q^3-q^2+8q-14$ for $q \ge 7$

(Barlotti 1965)

$m_2(4,q) < q^3-cq^{5/2}$ for $q > q_0(c)$ where $c < 1/4$

(Segre 1967)

$m_2(4,q) \le q^3-q^{3/2}+\frac{1}{2}q - \frac{1}{8}\sqrt{q}+1$

(Hill 1978a)

$m_2(4,q) \le q^3 - 8$ for $q=5$

(Hill 1978a)

$m_2(4,q) < q^3 - \frac{1}{4}q^{5/2} + 3q^2$ for $q > 121$

(Hirschfeld 1983a)

(ii) d=4, q even, q>2

$m_2(4,q) \le q^3$

(Tallini 1956)

$m_2(4,q) \le q^3-1$

(Barlotti 1965)

$m_2(4,q) < q^3 - \frac{1}{2}q^{3/2} + \frac{3}{4}q+1$

(Segre 1967)

(iii) d≥5, q odd

$m_2(d,q) \le q^{d-1}$

(Tallini 1956)

$m_2(d,q) \le q^{d-1} - (q-5)\theta_{d-4} + 1$ for $q \ge 7$

(Barlotti 1957)

$m_2(d,q) \le q^{d-1} - 2\theta_{d-4} + 1$ for $q=5$

$= \frac{1}{2}(49.5^{d-3} + 3)$

(Barlotti 1957)

$m_2(d,q) \le q^{d-1} - q^{d-2} + 20q^{d-3} - 1$

(Segre 1959)

$m_2(d,3) \le (3^{d+1}+23)/10$

(Bose and Srivastava 1964)

$m_2(d,q) \le q^{d-1} - q^{d-2} + 8q^{d-3} - 6\theta_{d-4} - 8$ for $q \ge 7$

(Barlotti 1965)

$m_2(d,q) \le q^{d-1} - \theta_{d-3} + 1$ for $q=7$

$= (41.7^{d-2}+7)/6$

(Barlotti 1965)

$m_2(d,3) \le 2.3^{d-2} + 2$

(Hill 1978a)

$m_2(d,5) \le 5^{d-1} - \frac{19}{2} 5^{d-4} + 3/2$

(Hill 1978a)

$m_2(d,7) \le 7^{d-1} - \frac{19}{2} 7^{d-4}(\frac{3}{4}\sqrt{7} - \frac{1}{3}) + 4/3$

(Hill 1978a)

$m_2(d,q) \le q^{d-1} - q^{d-2} + 8q^{d-3} - 15q^{d-4} - 2\theta_{d-5} + 1$

for $q>7$

(Hill 1978a)

$m_2(d,q) < q^{d-1} - cq^{d-3/2}$ for $q > q_0(c)$ where $c<1/4$

(Segre 1967)

$m_2(d,q) < q^{d-1} - \frac{1}{4}q^{d-3/2} + 3q^{d-2}$ for $q>121$

(Hirschfeld 1983a)

(iv) <u>$d \ge 5$, q even, $q>2$</u>

$m_2(d,q) \le q^{d-1}$

(Tallini 1956)

$m_2(d,q) \le q^{d-1} - \theta_{d-4} + 1$

(Barlotti 1957)

$$m_2(d,q) \le \{q^2-2q-1+(q-2)\theta_d + [(q-2)^2\theta_d^2+2(q^3-q-2)\theta_d+(q^2-2q-1)^2]^{\frac{1}{2}}\} \div \{2(q^2-q-1)\}$$ for all $q>2$

$\sim \quad q^{d-1} - q^{d-4}$ for large q

(Bose and Srivastava 1964)

$m_2(d,q) \le \{q^2-2q-1+(q-2)\theta_d\}/(q^2-q-1)$

(Barlotti 1965)

$m_2(d,q) \le q^{d-1} - 2\theta_{d-4} - 1$

(Barlotti 1965)

$m_2(d,q) < q^{d-1} - \frac{1}{2}q^{d-5/2} + \frac{3}{4}q\theta_{d-4} + 2$

(Segre 1967)

$m_2(d,q) < q^{d-1} - \frac{1}{2}q^{d-5/2} + \frac{3}{4}q^{d-3} - 2\theta_{d-5} + 1$

(Hill 1978a)

$m_2(d,q) < q^{d-1} - \frac{1}{2}q^{d-5/2} + \frac{3}{4}q^{d-3} - \frac{5}{4}\theta_{d-4} + 9/4$ for $q=2^{2e}$

(Thas 1968b)

$m_2(d,q) < q^{d-1} + (6-8\sqrt{2})\theta_{d-4} + 1$ for $q=8$

(Thas 1968b)

(c) *Lower Bounds*

$m_2(3t,q) \ge (q^{2t+2}-1)/(q^2-1)$

(Segre 1959)

$m_2(3t+1,q) \ge (2q^{2t+2}-q^{2t}-1)/(q^2-1)$

(Segre 1959)

$m_2(3t+1,q) \ge 2(q^{2t+2}-1)/(q^2-1)$

(Mukhopadhyay 1978)

$m_2(3t+2,q) \ge (q^{2t+3}-q^{2t+2}-q^{2t+1}-1)/(q^2-1)$ for q odd

(Segre 1959)

$m_2(3t+2,q) \ge (q^{2t+2}-1)/(q-1)$ for q odd

(Mukhopadhyay 1978)

$m_2(3t+2,q) \ge (q^{2t+3}+2q^{2t+2}-q^{2t+1}-q^{2t}-1)/(q^2-1)$ for q even

(Segre 1959)

$m_2(3t+2,q) \ge (q+2)(q^{2t+2}-1)/(q^2-1)$ for q even

(Mukhopadhyay 1978)

§4 r=n>2, s=1, d>2: (k;n)-caps

$m_n(d,q) \le n + (m_n(2,q)-n)\theta_{d-2}$

$m_3(3,q) \le 2q^2$ for $q>2$

(Bramwell and Wilson 1973)

$m_3(3,4) = 31$

(Hill 1978c)

$m_3(3,q) \le 2q^2+1 - d(q)$,

where $d(q)$ is the smallest positive integer d satisfying

$$d^2\left[\frac{2q^2-d}{q+1-d} + q^2 - q - 1\right] + d\left[\frac{q(2q^2-d)}{q+1-d} + 3q^3 - 6q\right] + 6q^3 - 2q^4 \geq 0$$

(Hill 1978c)

§5 r=d, s=d-1, d>2: k-arcs

$m(d,q) = d+2$ for $q \leq d+1$

(Bush 1952)

$m(3,q) = q+1$ for

(a) q odd, $q>3$

(Segre 1955b)

(b) q even, $q>2$

(Casse 1969, Gulati and Kounias 1970)

$m(4,q) = q+1$ for

(a) q odd, $q>3$

(Segre 1955b)

(b) q even, $q>4$

(Casse 1969, Gulati and Kounias 1970)

$m(d,q) = q+1$ for q odd and

(a) $q>(4d-5)^2$

(Thas 1968a)

(b) $f(q)>d-3$ (see §1 (10))

(Thas 1968a)

(c) $d = q-2$, $q-3$ or $q-4$

(Thas 1969a)

(d) $q-4>d>q - \frac{1}{4}\sqrt{q} - \frac{9}{4}$

(Thas 1969a)

$m(d,q) = q+1$ for q even and $d = q-3$ or $q-4$

(Thas 1969a)

$m(q-2,q) = q+2$

(Thas 1969a)

$\max(d+2,q+1) \leq m(d,q) \leq \max(d+2, q+d-3)$ for $d \geq 4$ and

(a) q odd

(Segre 1955b)

(b) q even

(Casse 1969)

(1) $m_2(d,q) = d+2$ for $q \le d+1$, and a $(d+2)$-arc is projectively unique.

Proof: $K = \{U_0, U_1, \ldots, U_d, U\}$ is a $(d+2)$-arc and any other is projectively equivalent to K. Let $P = P(a_0, \ldots, a_d)$ and $K' = K \cup \{P\}$. If $a_i=0$, then $K' \setminus \{U, U_i\}$ is a set of $d+1$ points in the prime with equation $x_i=0$. If no a_i is zero, then the $d+1$ coordinates of P can take any one of $q-1$ values. Since $q \le d+1$, so $q-1<d+1$. Hence $a_i=a_j$ for some $i \ne j$. So $K' \setminus \{U_i, U_j\}$ is a set of $d+1$ points in the prime with equation $x_i=x_j$. So K' is not a $(d+3)$-arc.

(2) A $(q+1)$-arc in $PG(3,q)$, q odd, is a twisted cubic C; that is, C is projectively equivalent to $\{P(1,t,t^2,t^3) \mid t \in GF(q) \cup \{\infty\}\}$

(Segre 1955b)

(3) A $(q+1)$-arc in $PG(3,q)$, $q = 2^h$, is projectively equivalent to $D(2^n)$ with $(n,h)=1$, where $D(e) = \{P(1,t,t^e,t^{e+1}) \mid t \in GF(q) \cup \{\infty\}\}$.

(Casse and Glynn 1982)

(4) A $(q+1)$-arc in $PG(d,q)$, q odd and (a) $q>(4d-5)^2$ or (b) $f(q)>d-3$, is a normal rational curve.

(Thas 1968a)

(5) A k-arc in $PG(d,q)$, q odd and

(a) $q+1 \ge k > q - \frac{1}{4}\sqrt{q} + d - \frac{1}{4}$ or

(b) $q+1 \ge k > q - f(q) + d - 2$,

is contained in a normal rational curve.

(Thas 1968a)

(6) When $k \ge d+3$, a k-arc exists in $PG(d,q)$ if and only if a k-arc exists in $PG(k-d-2, q)$.

(Thas 1969a, Halder and Heise 1974)

(7) In $PG(q-2,q)$, q even, a $(q+2)$-arc can be constructed by adding a point to a normal rational curve.

(Thas 1969b)

(8) From above,

$m(5,q) = q+1$ for $q>225$ and odd

$m(5,q) = q+1$ for $q = 7,8,9$

$m(5,q) \le q+2$ for $q>4$.

So, for q odd and $11 \le q \le 223$, is

$$m(5,q) = q+1 \quad \text{or} \quad q+2?$$

(9) From (4), a $(q+1)$-arc in $PG(4,q)$, $q>121$ and odd, is a normal rational curve. From (4)(b), a sufficient condition is that $f(q)>1$. Now, by §1, $f(9) = 1$. Recently, Casse and Glynn have given an example of a 10-arc K in $PG(4,9)$ which is not a normal rational curve. $K = \{U_0, U_1, U_2, U_3, U_4, P(1,1,\sigma^5,\sigma^7,\sigma^5), P(\sigma^5,1,1,\sigma^5,\sigma^7), P(\sigma^7,\sigma^5,1,1,\sigma^5), P(\sigma^5,\sigma^7,\sigma^5,1,1), P(1,\sigma^5,\sigma^7,\sigma^5,1)\}$; here $\sigma^2 = \sigma+1 \pmod 3$. This means that the matrix whose rows are the coordinate vectors of the last five points has *every* sub-determinant non-zero. Thus in (4) some numerical condition is always required.

§6 $r = n>d,\ s = d-1,\ d>2$: $(k;n)$-arcs

$$m(n,d-1;d,q) \le (n-d+1)q+n$$

$$m(n,d-1;d,q) \le (n-d+1)q+n-2 \quad \text{for} \quad q \not\equiv 0 \pmod{n-d+2}$$

(Barlotti 1965)

§7 $2<r<d,\ s = r-1$: $(k;r)$-sets

With $m = M_r(d,q)$,

$$q^{d+1} \ge 1 + \binom{m}{1}(q-1) + \ldots + \binom{m}{t}(q-1)^t \quad \text{for } r = 2t-1$$

$$q^{d+1} \ge 1 + \binom{m}{1}(q-1) + \ldots + \binom{m}{t}(q-1)^t + \binom{m-1}{t}(q-1)^{t+1} \quad \text{for } r = 2t$$

(Rao 1947, Bose 1947, 1961)

If k is the largest integer such that

$$\binom{k}{1}(q-1) + \binom{k}{2}(q-1)^2 + \ldots + \binom{k}{r}(q-1)^r < q^{d+1} - 1,$$

then

$$M_r(d,q) \ge k+1$$

(Gilbert 1952, Varshamov 1957)

$$M_r(nd,q) \ge M_r(d,q^n)$$

(Bose and Ray-Chaudhuri 1960, Peterson 1961)

$$M_r(d,q) = d+2 \quad \text{for} \quad q \le \frac{r+1}{d-r+1}$$

(Tallini 1960)

$M_r(d,q) \le d+2+t$ for $t \ge 0$ such that $r \ge 3q^t + q^{t-1} - 2$

$$\text{and} \quad r \ge \frac{(q-1)q^{t+1}(d+t+1) - q^t(q+1)+2}{q^{t+2} - 1}$$

(Tallini 1960)

$M_{r+1}(d+1,q) \le 1 + M_r(d,q)$

(Gulati 1971a)

$M_{d-1}(d,q) \le q^2 + d - 2$ for q odd

(Gulati 1971a)

$M_3(4,q) \le q(q-1)$ for $q \ge 4$

(Gulati and Kounias 1970)

$M_3(d,q) \le q^{d-2} - (q+1)\theta_{d-4} + 1$ for $d \ge 5$, $q \ge 4$

(Gulati and Kounias 1970)

$M_4(5,q) \le q^2 - q + 1$ for $q \ge 4$

(Gulati 1972)

$M_4(5,q) \le q^2 - 2q$ for q odd, $q \ge 5$

(Gulati 1972)

$M_4(d,q) \le q^{d-3} - (2q+1)\theta_{d-5} + 1$ for $d \ge 6$, q odd, $q \ge 5$

(Gulati 1972)

$M_r(d,q) = d+3$ for $d-1 \ge q > 2$ and $\frac{r+1}{d-r+1} < q \le g_q(r,d)$

where $g_q(r,d) = \dfrac{r+c'+\sqrt{\{(r+c')^2+4(r+1)(d-r+c'')\}}}{2(d-r"c")}$

with $d-r \equiv c \pmod{q+1}$ and

c	0	1	2 ...	q-3	q-2	q-1	q
c'	0	-1	-2 ...	-(q-3)	2	1	1
c"	1	1	1 ...	1	2	2	1

(Gulati and Kounias 1973)

$M_r(d,q) = d+n$ where, for $n \ge 3$,

(a) $d-r = (r_1+1)\theta_{n-2} - r_0$ such that $r_1 \ge 0$ and $1 \le r_0 \le \theta_{n-2}$;

(b) $n-1 \le r_0 \le n+q-2$;

(c) $T_n(d-r,q) = (r_1+1)q^{n-1} - \{r_0-(n-1)\}q-1$;

(d) $T_n(d-r,q) < r+1 \le T_{n-1}(d-r,q)$.

(Hamada and Tamari 1976, 1978)

In the last result, if q, r and n are given, then conditions (a)-(d) define those values of d for which the result holds. In practice this is only going to work for small n. Given r,q and a possible d, then for each integer $n \ge 3$, condition (a) defines r_1 and r_0. If r_0 does not satisfy condition (b), then this value of n is already rejected.

Now, (c) calculates the function T_n. Finally, (d) tells us which value of n, if any, is suitable. It must be emphasised that this algorithm does not necessarily provide a solution for every r and d.

We give an example for $q = 5$ in which $n = 4$. Suppose $d-r = 63$. Then condition (a) with $n = 4$ and $n = 3$ gives

$$63 = 31r_1 + 31 - r_0 = 6r_1' + 6 - r_0'.$$

Hence

$$r_1 = 2,\ r_0 = 30,\ r_1' = 10,\ r_0' = 3.$$

So

$$T_4 = (r_1 + 1)5^3 - (r_0 - 3)\ 5 - 1 = 239,$$
$$T_3 = (r_1' + 1)5^2 - (r_0' - 2)5 - 1 = 269.$$

Thus, providing that

$$T_4 = 239 < r + 1 \le 269 = T_3,$$

the required value of n is 4. Hence

$$M_{239}(302,5) = 306,$$
$$M_{240}(303,5) = 307,$$
$$\vdots$$
$$M_{268}(331,5) = 335.$$

(1) For the above bound, corresponding $(d+n;r)$-sets are constructed.

(2) For $r \ge 4$, a $(k;r)$-set is the intersection of complete sets of kind two (k'-caps).

(Tallini 1961)

(3) The only $(k;r)$-sets not contained in a $(k';2)$-set are projectively equivalent to K_1 or K_2. Here $K_1 = \{U_0,U_1,U_2,U_3,U\} \subset PG(3,2)$; that is, K_1 is a normal rational curve, which is a $(5;3)$-set.
$K_2 = \{U_0,U_1,U_2,U_3,U_4,U,P(0,1,2,2,1),P(1,0,2,1,2),P(1,1,0,2,2),$ $P(1,2,2,0,1),P(1,2,1,2,0)\} \subset PG(4,3)$ and is an $(11;3)$-set.

(Tallini 1961, Makowski 1963, Migliori 1976)

(a) $q = 2$

$$M_3(4,2) = 6$$
$$M_3(5,2) = 8$$
$$M_3(6,2) = 11$$

(Rao 1947)

$$M_r(d,2) = d+2 \quad \text{for} \quad r \ge (2d+1)/3$$

(Tallini 1960)

$M_3(7,2) = 17$

$M_3(d,2) \leq 3.2^{d-5} + 5$ for $d\geq 8$

(Seiden 1973)

$M_4(6,2) = 9$

(Gulati 1971b)

$M_r(d,2) = d+3$ for $f_2(d) \leq r \leq \frac{2}{3}d$

$M_r(d,2) \geq d+4$ for $r<f_2(d)$

where $d-r \equiv c \pmod 3$ and $f_2(d) = \frac{1}{7}(4d+3+c')$

c	0	1	2
c'	0	2	1

(Gulati, Johnson and Koehn 1973)

$M_3(d+1,2) \geq M_3(d,2) + 2^{d-7}$ for $7\leq d\leq 11$

(Mukhopadhyay 1978)

$M_4(d+1,2) = M_3(d,2)+1$ for $d\geq 3$

(Mukhopadhyay 1978)

(b) $q = 3$

$M_r(d,3) = d+2$ for $r \geq (3d+2)/4$

(Tallini 1960)

$M_3(4,3) = 11$

(Gulati and Kounias 1970)

$M_{d-2}(d,3) = d+4$ for $7\leq d\leq 9$

(Gulati 1971a)

$M_4(5,3) = 12$

(Gulati 1971a)

$M_r(d,3) = d+3$ for $f_3(d) \leq r \leq (3d+1)/4$

$M_r(d,3) \geq d+4$ for $r<f_3(d)$

where $d-r \equiv c \pmod 4$ and $f_3(d) = \frac{1}{13}(9d+5+3c')$

c	0	1	2	3
c'	1	2	3	0

(Gulati 1971a)

$M_3(5,3) = 13$

(Hill 1978b)

$17\leq M_3(6,3)\leq 33$

(Hill 1978b)

$13\leq M_4(6,3)\leq 14$

(Hill 1978b)

(c) $q = 4$

$M_r(d,4) = d+2$ for $r \geq (4d+3)/5$

(Tallini 1960)

$M_r(d,4) = d+3$ for $f_4(d) \leq r \leq (4d+2)/5$

$M_r(d,4) \geq d+4$ for $r < f_4(d)$

where $d-r \equiv c \pmod 5$ and $f_4(d) = (16d+13+c')/21$.

c	0	1	2	3	4
c'	2	1	0	4	3

(Gulati, Johnson and Koehn 1973)

REFERENCES

Several works, not referred to in the text, are included. They are Haimulin 1966, Maneri and Silverman 1966, 1971.

Barlotti, A. (1955). Un' estensione del teorema di Segre-Kustaanheimo. Boll. Un. Mat. Ital. 10, 96-98.

Barlotti, A. (1956). Su {k;n}-archi di un piano lineare finito. Boll. Un. Mat. Ital. 11, 553-556.

Barlotti, A. (1957). Un limitazione superiore per il numero dei punti appartenenti a una k-calotta C(k,0) di uno spazio lineare finito. Boll. Un. Mat. Ital. 12, 67-70.

Barlotti, A. (1965). Some topics in finite geometrical structures. Institute of Statistics mimeo series no. 439, Univ. of North Carolina.

Bose, R.C. (1947). Mathematical theory of the symmetrical factorial design. Sankyhā 8, 107-166.

Bose, R.C. (1959). On the application of finite projective geometry for deriving a certain series of balanced Kirkman arrangements. Golden Jubilee Commemoration Volume (1958-9) Calcutta Math. Soc., 341-354.

Bose, R.C. (1961). On some connections between the design of experiments and information theory. Bull. Inst. Internat. Statist. 38, 257-271.

Bose, R.C. & Ray-Chaudhuri, D.K. (1960a). On a class of error correcting binary group codes. Inform. and Control 3, 68-79.

Bose, R.C. & Ray-Chaudhuri, D.K. (1960b). Further results on error correcting binary group codes. Inform. and Control 3, 279-290.

Bose, R.C. & Srivastava, J.N. (1964). On a bound useful in the theory of factorial designs and error correcting codes. Ann. Math. Statist. 35, 408-414.

Bramwell, D.L. & Wilson, B.J. (1973). Cubic caps in three dimensional Galois space. Proc. Royal Irish Acad. 73A, no.20, 279-283.

Bush, K.A. (1952). Orthogonal arrays of index unity. Ann. Math. Statist. 23, 426-434.

Casse, L.R.A. (1969). A solution to Beniamino Segre's "Problem $I_{r,q}$" for q even. Atti Accad. Naz. Lincei Rend. 46, 13-20.

Casse, L.R.A. & Glynn, D.G. (1982). The solution to Beniamino Segre's problem $I_{r,q}$, $r = 3$, $q = 2^h$. Geom. Dedicata 13, 157-164.

Cossu, A. (1961). Su alcune proprietà dei {k,n}-archi di un piano proiettivo sopra un corpo finito. Rend. Mat. e Appl. 20, 271-277.

Denniston, R.H.F. (1969). Some maximal arcs in finite projective planes. J. Comb. Theory 6, 317-319.

Gilbert, E.N. (1952). A comparison of signalling alphabets. Bell System Tech. J. 31, 504-522.

Gulati, B.R. (1971a). On maximal (k,t)-sets. Ann. Inst. Statist. Math. 23, 279-292; correction 23, 527-529.

Gulati, B.R. (1971b). On orthogonal arrays of strength five. Trabajos Estadist. 23, 51-77.

Gulati, B.R. (1972). More about maximal (n,r)-sets. Inform. and Control 20, 188-191.

Gulati, B.R., Johnson, B. McK. & Koehn, U. (1973). On maximal t-linearly independent sets. J. Combin. Theory Ser. A 15, 45-53.

Gulati, B.R. & Kounias, E.G. (1970). On bounds useful in the theory of symmetrical factorial designs. J. Roy. Statist. Soc. Ser. B 32, 123-133.

Gulati, B.R. & Kounias, E.G. (1973). Maximal sets of points in finite projective spaces, no t linearly dependent. J. Combin. Theory Ser. A 15, 54-65.

Haimulin, J.N. (1966). Some properties of $\{k;n\}_q$-arcs in Galois planes. Soviet Math. Dokl. 7, 1100-1103.

Halder, H.-R. & Heise, W. (1974). On the existence of finite chain-m-structures and k-arcs in finite projective space. Geom. Dedicata 3, 483-486.

Hall, M. (1975). Ovals in the Desarguesian plane of order 16. Ann. Mat. Pura Appl. 102, 159-176.

Hamada, N. & Tamari, F. (1976). Construction of maximal t-linearly independent sets. Essays in Probability and Statistics (Editors S. Ikeda et al.) Keibundo Matsumoto Printing Co., pp.41-55.

Hamada, N. & Tamari, F. (1978). On a geometrical method of construction of maximal t-linearly independent sets. J. Combin. Theory Ser. A 25, 14-28.

Hill, R. (1973). On the largest size of cap in $S_{5,3}$. Atti. Accad. Naz. Lincei Rend. 54, 378-384.

Hill, R. (1978a). Caps and codes. Discrete Math. 22, 111-137.

Hill, R. (1978b). Packing problems in Galois geometries over GF(3). Geom. Dedicata 7, 363-373.

Hill, R. (1978c). Some results concerning linear codes and (k,3)-caps in three-dimensional Galois space. Math. Proc. Cambridge Philos. Soc. 84, 191-205.

Hill, R. (1983a). On Pellegrino's 20-caps in $S_{4,3}$. Combinatorial Geometries and their Applications (Rome 1981), Ann. Discrete Math. 18, 443-448.

Hill, R. (1983b). Some problems concerning (k,n)-arcs in finite projective planes. Combinatorial and Incidence Geometry (Bolzano 1982). To appear.

Hill, R. & Mason, J.R.M. (1981). On (k,n)-arcs and the falsity of the Lunelli-Sce conjecture. Finite geometries and designs, London Math. Soc. Lecture Note Series 49, Cambridge University Press, 153-168.

Hirschfeld, J.W.P. (1979). Projective geometries over finite fields. Oxford University Press.

Hirschfeld, J.W.P. (1983a). Caps in elliptic quadrics. Combinatorial Geometries and their Applications (Rome 1981), Ann. Discrete Math. 18, 449-466.

Hirschfeld, J.W.P. (1983b). The Weil conjectures in finite geometry. Tenth Australian Conference on Combinatorial Mathematics (Adelaide 1982). To appear.

Korchmáros, G. (1978). Gruppi di collineazioni transitivi sui punti di un ovale [(q+2)-arco] di $S_{2,q}$, q pari. Atti Sem. Mat. Fis. Univ. Modena 27, 89-105.

Lunelli, L. & Sce, M. (1964). Considerazione aritmetiche e risultati sperimentali sui $\{K;n\}_q$-archi. Ist. Lombardo Accad. Sci. Rend. A 98, 3-52.

MacWilliams, F.J. & Sloane, N.J.A. (1977). The theory of error-correcting codes. North Holland.

Makowski, A. (1963). Remarks on a paper of Tallini. Acta Arith. 8, 469-470.

Maneri, C. & Silverman, R. (1966). A vector-space packing problem. J. Algebra 4, 321-330.

Maneri. C. & Silverman, R. (1971). A combinatorial problem with applications to geometry. J. Combin. Theory Ser. A 11, 118-121.

Mason, J.R.M. (1982). On the maximum sizes of certain (k,n)-arcs in finite projective geometries. Math. Proc. Cambridge Philos. Soc. 91, 153-170.

Migliori, G. (1976). Calotte di specie s in uno spazio r-dimensionale di Galois. Atti. Accad. Naz. Lincei Rend. 60, 789-792.

Mukhopadhyay, A.C. (1978). Lower bounds on $m_t(r,s)$. J. Combin. Theory Ser. A 25, 1-13.

Panella, G. (1955). Caratterizzazione delle quadriche di uno spazio (tridimensionale) lineare sopra un corpo finito. Boll. Un. Mat. Ital. 10, 507-513.

Pellegrino, G. (1970). Sul massimo ordine delle calotte in $S_{4,3}$. Matematiche (Catania) 25, 1-9.

Pellegrino, G. (1981). Archi completi di ordine (q+3)/2 nei piani di Galois, $S_{2,q}$, con $q \equiv 3 \pmod 4$. Rend. Circ. Mat. Palermo 30, 311-320.

Pellegrino, G. (1983). Archi completi, di ordine (q+3)/2, nei piani di Galois, $S_{2,q}$, con $q \equiv 1 \pmod 4$. To appear.

Peterson, W.W. (1961). Error-correcting codes. M.I.T. Press.

Qvist, B. (1952). Some remarks concerning curves of the second degree in a finite plane. Ann. Acad. Sci. Fenn. Ser. A, no. 134.

Rao, C.R. (1947). Factorial experiments derivable from combinatorial arrangements of arrays. J. Roy. Statist. Soc., Suppl. 9, 128-139.

Segre, B. (1954). Sulle ovali nei piani lineari finiti. Atti Accad. Naz. Lincei Rend. 17, 1-2.

Segre, B. (1955a). Ovals in a finite projective plane. Canad. J. Math. 7, 414-416.

Segre, B. (1955b). Curve razionali normali e k-archi negli spazi finiti. Ann. Mat. Pura. Appl. 39, 357-379.

Segre, B. (1957). Sui k-archi nei piani finiti di caratteristica due. Rev. Math. Pures Appl. 2, 289-300.

Segre, B. (1959). Le geometrie di Galois. Ann. Mat. Pura Appl. 48, 1-97.

Segre, B. (1960). Lectures on modern geometry (with an appendix by L. Lombardo-Radice). Cremonese.

Segre, B. (1967). Introduction to Galois geometries. Atti. Accad. Naz. Lincei Mem. 8, 133-236.

Seiden, E. (1950). A theorem in finite projective geometry and an application to statistics. Proc. Amer. Math. Soc. 1, 282-286.

Seiden, E. (1973). On the problem of construction and uniqueness of saturated 2_R^{k-p} designs. A survey of Combinatorial Theory (Colorado State University, 1971), North Holland, 397-401.

Tallini, G. (1956). Sulle k-calotte di uno spazio lineare finito. Ann. Mat. Pura Appl. 42, 119-164.

Tallini, G. (1960). Le geometrie di Galois e le loro applicazioni alla statistica e alla teoria delle informazioni. Rend. Mat e Appl. 19, 379-400.

Tallini, G. (1961). On caps of kind s in a Galois r-dimensional space. Acta Arith. 7, 19-28.

Tallini, G. (1963). Un 'applicazione delle geometrie di Galois a questioni di statistica. Atti. Accad. Naz. Lincei Rend. 35, 479-485.

Thas, J.A. (1968a). Normal rational curves and k-arcs in Galois spaces. Rend. Mat. 1, 331-334.

Thas, J.A. (1968b). Contribution to the theory of k-caps in projective Galois spaces. (Flemish) Med. Konink. Vlaamse Acad. Wetensch. 30, No. 13.

Thas, J.A. (1969a). Connection between the Grassmannian $G_{k-1;n}$ and the set of the k-arcs of the Galois space $S_{n,q}$. Rend. Mat. 2, 121-134.

Thas, J.A. (1969b). Normal rational curves and (q+2)-arcs in a Galois space $S_{q-2,q}$ ($q = 2^h$). Atti Accad. Naz. Lincei Rend. 47, 249-252.

Thas, J.A. (1975). Some results concerning {(q+1)(n-1);n}-arcs and {(q+1)(n-1) + 1;n}-arcs in finite projective planes of order q. J. Combin. Theory Ser. A 19, 228-232.

Tits, J. (1962). Ovoides et groupes de Suzuki. Arch. Math. 13, 187-198.

Varshamov, R.R. (1957). Estimate of the number of signals in error correcting codes. Dokl. Akad. Nauk SSSR 117, 739-741.

Waterhouse, W.C. (1969). Abelian varieties over finite fields. Ann. Sci. École Norm. Sup. 2, 521-560.

Wilson, B.J. (1974). A note on (k,n)-arcs. Proc. Cambridge Philos. Soc. 76, 57-59.

Wilson, B.J. (1982). Incompleteness of (nq+n-q-2,n)-arcs in finite projective planes of even order. Math. Proc. Cambridge Philos. Soc. 91, 1-8.

QUASIGROUP IDENTITIES AND ORTHOGONAL ARRAYS

C. C. Lindner
Department of Mathematics, Auburn University,
Auburn, Alabama 36849 U.S.A.

1. Introduction. Let (Q,o) be a quasigroup of order n and define an $n^2 \times 3$ array A by (x,y,z) is a row of A if and only if $x \circ y = z$. As a consequence of the fact that the equations $a \circ x = b$ and $y \circ a = b$ are uniquely solvable for all $a,b \in Q$, if we run our fingers down any two columns of A we get each ordered pair belonging to $Q \times Q$ exactly once. An $n^2 \times 3$ array with this property is called an orthogonal array and it doesn't take the wisdom of a saint to see that this construction can be reversed. That is, if A is any $n^2 \times 3$ orthogonal array (defined on a set Q) and we define a binary operation o on Q by $x \circ y = z$ if and only if (x,y,z) is a row of A, then (Q,o) is a quasigroup. Hence we can think of a quasigroup as an $n^2 \times 3$ orthogonal array and conversely. Now given an $n^2 \times 3$ orthogonal array A there is the irresistable urge to permute the columns of A. One of the reasons for this urge is that the resulting $n^2 \times 3$ array is still an orthogonal array. (If running our fingers down any two columns of A gives every ordered pair of $Q \times Q$ exactly once, the same must be true (of course) if we rearrange the columns.) If $\alpha \in S_3$ (the symmetric group on $\{1,2,3\}$) and A is an $n^2 \times 3$ orthogonal array we will denote by $A\alpha$ the orthogonal array obtained by permuting the columns of A according to α. Two orthogonal arrays are equal if and only if they define the same quasigroup. In other words, if we disregard the level at which the rows occur, the two orthogonal arrays contain exactly the same rows.

Two $n^2 \times 3$ orthogonal arrays A and B are said to be conjugate provided there is at least one $\alpha \in S_3$ such that $A\alpha = B$. Two quasigroups are said to be conjugate provided their corresponding orthogonal arrays are conjugate.

<u>Example 1.1.</u>

o_1	1	2	3	4
1	1	3	4	2
2	4	2	1	3
3	2	4	3	1
4	3	1	2	4

o_2	1	2	3	4
1	1	4	2	3
2	3	2	4	1
3	4	1	3	2
4	2	3	1	4

o_3	1	2	3	4
1	1	3	4	2
2	4	2	1	3
3	2	4	3	1
4	3	1	2	4

1	1	1
1	2	3
1	3	4
1	4	2
2	1	4
2	2	2
2	3	1
2	4	3
3	1	2
3	2	4
3	3	3
3	4	1
4	1	3
4	2	1
4	3	2
4	4	4

A

1	1	1
2	1	3
3	1	4
4	1	2
1	2	4
2	2	2
3	2	1
4	2	3
1	3	2
2	3	4
3	3	3
4	3	1
1	4	3
2	4	1
3	4	2
4	4	4

A(12)

1	1	1
3	1	2
4	1	3
2	1	4
4	2	1
2	2	2
1	2	3
3	2	4
2	3	1
4	3	2
3	3	3
1	3	4
3	4	1
1	4	2
2	4	3
4	4	4

A(123)

Note that in the above example $A = A(123)$; i.e., A is invariant under conjugation by (123). It follows at once that A must be invariant under conjugation by $\langle(123)\rangle = \{(1), (123), (132)\}$ the subgroup of S_3 generated by (123). A quick check shows that A is *not* invariant under any of the remaining three permutations (12), (13), and (23) of S_3; and so *the* set of all permutations of S_3 which fix A is precisely the cyclic group $\langle(123)\rangle$ of S_3.

In general, the *conjugate invariant subgroup* H of an orthogonal array B is defined by

$$H = \{\alpha \in S_3 \mid B\alpha = B\}.$$

We remark that this definition requires that H consists of *all* permutations which fix B. That is to say, H is not the conjugate invariant subgroup of B if H is contained properly in a subgroup of S_3 which fixes B. Now if we look carefully at the quasigroup (Q,o_1) associated with the orthogonal array A in Example 1.1 we will see that (Q,o_1) satisfies the identity $(xy)x = y$, (and also the identity $x(yx) = y$, which is not surprizing since the identities $(xy)x = y$ and $x(yx) = y$ are equivalent). This is not happenstance since, in general, a quasigroup satisfies the identity $(xy)x = y$ if and only if its associated orthogonal array is invariant under conjugation by $C_3 = \langle(123)\rangle$. This is quite easy to see: Let (Q,o) be a quasigroup and B its associated orthogonal array. Suppose (Q,o) satisfies $(xy)x = y$. If $(x,y,z) \in B$ then $x \circ y = z$ implies $(x \circ y) \circ x = y$ implies $(z,x,y) \in B$ which implies that B is *at least* invariant under conjugation by $C_3 = \langle(123)\rangle$. On the other hand suppose B is invariant under conjugation by $C_3 = \langle(123)\rangle$. If $x \circ y = z$, then $(x,y,z) \in B$ implies $(x \circ y) \circ x = z \circ x = y$ and so (Q,o) satisfies the identity $(xy)x = y$. Quasigroups which satisfy the identity $(xy)x = y$ (and therefore also satisfy the identity $x(yx) = y$) are called *semisymmetric*. The vernacular is certainly well taken, since a semisymmetric quasigroup is invariant under conjugation by at least half of the symmetric group S_3, namely $C_3 = \langle(123)\rangle$. It should now come as no surprise that a quasigroup which is invariant under conjugation by the full symmetric group S_3 is called *totally symmetric*. These quasigroups can be characterized by the two identities $x(xy) = y$ and $(yx)x = y$ (which, by the way, imply the commutative law $xy = yx$).

The above observations are a good illustration of the theme of this set of notes...the interaction between orthogonal arrays (both $n^2 \times 3$ and $n^2 \times 4$) and quasigroup identities. Rather than give a summary of what follows, we plunge right in. Things will become clear as we progress.

2. $n^2 \times 3$ orthogonal arrays and identities. Let A be an $n^2 \times 3$ orthogonal array and (Q,o) the corresponding quasigroup. The following two tables are self explanatory. All quantification is with respect to the five identities $xy = yx$, $(yx)x = y$, $x(xy) = y$, $(xy)x = y$, and $x(yx) = y$.

A invariant under	(Q,o) satisfies
(12)	$xy = yx$
(13)	$(yx)x = y$
(23)	$x(xy) = y$
(123)	$(xy)x = y$
(132)	$x(yx) = y$

Table 2.1

Conjugate invariant subgroup of A is	(Q,o) satisfies
<1>	none of the identities
<(12)>	$xy = yx$ only
<(13)>	$(yx)x = y$ only
<(23)>	$x(xy) = y$ only
<(123)>	$(xy)x = y \equiv x(yx) = y$ only
S_3	all five identities

Table 2.2

It is a straightforward exercise to verify the information in the above tables. A few comments concerning vernacular are in order. Any quasigroup satisfying $xy = yx$ is called commutative, any quasigroup satisfying the equivalent identities $(xy)x = y$ and $x(yx) = y$ is called semisymmetric, and any quasigroup satisfying all five identities is called totally symmetric. Since the identities $(xy)x = y$ and $x(yx) = y$ are equivalent we need only one to describe a semisymmetric quasigroup and we will generally use $x(yx) = y$. It is immediate that any two of the identities $xy = yx$, $x(xy) = y$, $(yx)x = y$, and $x(yx) = y$ imply all five identities, since the two corresponding permutations generate S_3. In what follows we will generally use the two identities $x(xy) = y$ and $(yx)x = y$ to describe a totally symmetric quasigroup. (We could, of course, just as well use $xy = yx$ and $x(xy) = y$, for example).

3. The spectrum for subgroups of S_3. For each subgroup H of S_3, the spectrum of H is the set of all n such that there is a quasigroup of order n (= $n^2 \times 3$ orthogonal array) having H as its conjugate invariant subgroup. Recall, that H is the conjugate invariant subgroup of a quasigroup (Q,o) if and only if H is the largest subgroup of S_3 which fixes (Q,o). The following table is a summary of the results in [9]. All quantification is with respect to the identities $I = \{xy = yx, x(xy) = y, (yx)x = y, x(yx) = y \equiv (xy)x = y\}$.

subgroup of S_3	spectrum	(Q,o) belongs to the spectrum iff it satisfies the identities	Comments
$\langle 1 \rangle$	all n except 1, 2, and 3	None	None
$\langle (12) \rangle$	all n except 1 and 2	$xy = yx$ only	There are, of course, commutative quasigroups of orders 1 and 2 <u>but</u> they are totally symmetric.
$\langle (13) \rangle$	all n except 1 and 2	$(yx)x = y$ only	see above
$\langle (23) \rangle$	all n except 1 and 2	$x(xy) = y$ only	see above
$\langle (123) \rangle$	all n except 1, 2, and 3	$x(yx) = y \equiv (xy)x = y$ only	There are semi-symmetric quasigroups of orders 1, 2, 3 <u>but</u> they are all totally symmetric.
S_3	all n	$x(xy) = y$ and $(yx)x = y$	Any two of the identities $xy = yx$, $x(yx) = y$, $x(xy) = y$, and $(yx)x = y$ imply all four. The author prefers, for some strange reason, the two identities $x(xy) = y$ and $(yx)x = y$.

Table 3.1

We verify the table line by line.

<u>$\langle 1 \rangle$</u>. It is not possible to construct a quasigroup of order 1, 2, or 3 which does not satisfy at least one of the identities in I (this is an easy exercise) and so 1, 2, and 3 cannot belong to the spectrum for $\langle 1 \rangle$. Now let (Q,o) be the quasigroup of order 4 given by the accompanying table.

o	1	2	3	4
1	1	2	3	4
2	2	3	4	1
3	4	1	2	3
4	3	4	1	2

Trivially, (Q,o) satisfies none of the identities in I and so 4 belongs to the spectrum for <1>. In [5] Trevor Evans has shown that a (partial) quasigroup of order n can be embedded in a quasigroup of order t for every $t \geq 2n$. Hence there is a quasigroup of order t satisfying none of the identities in I for every $t \geq 8$. The cases n = 5, 6, and 7 can be handled by examples.

<u><(12)>, <(13)>, <(23)></u>. To begin with, the identities xy = yx, (yx)x = y, and x(xy) = y are <u>conjugate</u>. That is, if (Q,o) is a quasigroup satisfying any <u>one</u> of the identities in K = {xy = yx,(yx)x = y, x(xy) = y} and we choose any one of the remaining two identities in K, then a suitable conjugate of (Q,o) satisfies this identity. Now if (Q,o) satisfies xy = yx <u>only</u> (with respect to I) and (Q, ⊗) is a conjugate of (Q,o) which satisfies x(xy) = y, then (Q, ⊗) must satisfy x(xy) = y only. Otherwise, the conjugate invariant subgroup of (Q, ⊗) would contain <(23)> properly which implies that (Q, ⊗) is totally symmetric which implies that (Q,o) and (Q, ⊗) are equal which implies that (Q,o) is totally symmetric which contradicts the fact that (Q,o) satisfies xy = yx only. Whew!

Now, let (Q,o) be the quasigroup of order 3 given by the accompanying table.

o	1	2	3
1	1	2	3
2	2	3	1
3	3	1	2

Then (Q,o) satisfies $xy = yx$ only. Allan Cruse [4] has shown that a diagonalized commutative quasigroup of order n can be embedded in a commutative (not necessarily diagonalized) quasigroup of order t for every $t \geq 6$. The cases $n = 4$ and 5 can be taken care of by example.

<u><(123)></u>. Any quasigroup of order 1, 2, or 3 which satisfies $x(yx) = y$ must also satisfy $xy = yx$ and so is totally symmetric. Hence we must begin with order 4. The accompanying quasigroups of orders 3, 4, and 5 are necessary for what follows.

o_1	1	2	3
1	1	3	2
2	3	2	1
3	2	1	3

(A,o_1)

o_2	∞_1	1	2	3
∞_1	∞_1	2	3	1
1	3	1	∞_1	2
2	1	3	2	∞_1
3	2	∞_1	1	3

(B,o_2)

o_3	∞_1	∞_2	1	2	3
∞_1	∞_1	∞_2	2	3	1
∞_2	∞_2	∞_1	3	1	2
1	3	2	1	∞_1	∞_2
2	1	3	∞_2	2	∞_1
3	2	1	∞_1	∞_2	3

(C,o_3)

A quick glance reveals that (A,o_1) is, of course, totally symmetric while (B,o_2) and (C,o_3) satisfy $x(yx) = y$ only. We can handle $n = 6, 7, 8, 9, 10$ and 11 by example and so we can assume $n \geq 12$. Now write $n = 3\cdot q + k$ ($k = 0, 1$, or 2), take $Q = \{1,2,3,\ldots,q\}$, and let $K = \emptyset$, $\{\infty_1\}$, or $\{\infty_1,\infty_2\}$ depending upon whether $k = 0, 1$, or 2. Finally, let (Q,o) be an idempotent quasigroup $(x^2 = x)$ which is not commutative (at least one pair $a \neq b$ such that $a \circ b \neq b \circ a$) and define a binary operation $\otimes$ on $S = K \cup (Q \times \{1,2,3\})$ by:

(1) For each $i \in Q$ define a copy of (A,o_1) on $\{(i,1), (i,2), (i,3)\}$ if $k = 0$, a copy of (B,o_2) on $\{\infty_1,(i,1),(i,2),(i,3)\}$ if $k = 1$, and a copy of (C,o_3) on $\{\infty_1, \infty_2, (i,1), (i,2), (i,3)\}$ if $k = 2$; and

(2) for each ordered pair (x,y), $x \neq y \in Q$, define $(x,1) \otimes (y,1) = (x \circ y,2)$, $(y,1) \otimes (x \circ y,2) = (x,1)$, $(x \circ y,2) \otimes (x,1) = (y,1)$; $(x,2) \otimes (y,2) = (x \circ y,3)$, $(y,2) \otimes (x \circ y,3) = (x,2)$, $(x \circ y,3) \otimes (x,2) = (y,2)$; $(x,3) \otimes (y,3) = (x \circ y,1)$, $(y,3) \otimes (x \circ y,1) = (x,3)$, and $(x \circ y,1) \otimes (x,3) = (y,3)$.

It is straightforward to see that $(S, \otimes)$ is a quasigroup satisfying $x(yx) = y$ and which is not commutative (since (Q,o) is not commutative). Hence there is a quasigroup of every order $n \geq 4$ satisfying the identity $x(yx) = x$ ($\equiv (xy)x = y$) only.

$\underline{S_3}$. This case is trivial. Define a binary operation $\otimes$ on Z_n by $x \times y = -(x + y) (\mathrm{mod}\ n)$. It is immediate that $(Z_n, \otimes)$ is totally symmetric.

4. The idempotent spectrum for subgroups of S_3.

The idempotent spectrum of the subgroup H of S_3 is the set of all n such that there is an idempotent quasigroup of order n with H as its conjugate invariant subgroup. The introduction of the idempotent law is a non-trivial restriction. All quantification is now with respect to the identities $E = \{x^2 = x,\ xy = yx,\ (yx)x = y,\ x(xy) = y,$
$x(yx) = y \equiv (xy)x = y\}$.

subgroup of S_3	idempotent spectrum	(Q,o) belongs to the idempotent spectrum iff it satisfies the identities.	Comments
⟨1⟩	all n except 1, 2, 3 and 4	$x^2 = x$ only	None
⟨(12)⟩	all odd $n \geq 5$	$x^2 = x$, $xy = yx$ only	There is only one idempotent quasigroup of order 1 or 3 and each is totally symmetric.
⟨(13)⟩	all odd $n \geq 5$	$x^2 = x$, $(yx)x = y$ only	see above
⟨(23)⟩	all odd $n \geq 5$	$x^2 = x$, $x(xy) = y$ only	see above
⟨(123)⟩	all $n \equiv 0$ or 1 (mod 3) except n = 1, 3 or 6.	$x^2 = x$, $x(yx) = y$ only	Any quasigroup satisfying $x^2 = x$ and $x(yx) = y$ is called a Mendelsohn quasigroup. Such quasigroups exist for orders 1 and 3 but are totally symmetric. No such quasigroup exists of order 6.
S_3	all $n \equiv 1$ or 3 (mod 6)	$x^2 = x$ idempotent; $x(xy) = y$, $(yx)x = y$ Steiner identities	An idempotent totally symmetric quasigroup is called a Steiner quasigroup. Note: Every Steiner quasigroup is Mendelsohn but the converse is not necessarily true.

Table 4.1

As before we verify the table line by line.

<u><1></u>. There is exactly one idempotent quasigroup of order 1 and 3, none of order 2, and two of order 4; and each of these idempotent quasigroups satisfies at least one identity in E other than $x^2 = x$. Let (Q,o) be the idempotent quasigroup of order 5 given below. Then (Q,o) satisfies only the identity $x^2 = x$ belonging to E.

(Q,o) =

o	1	2	3	4	5
1	1	3	5	2	4
2	5	2	4	3	1
3	4	1	3	5	2
4	2	5	1	4	3
5	3	4	2	1	5

A.J.W. Hilton [7] has shown that an idempotent quasigroup of order n can be embedded in an idempotent quasigroup of order t for <u>every</u> $t \geq 2n + 1$; and so embedding (Q,o) in an idempotent quasigroup of order t for every $t \geq 11$ gives an idempotent quasigroup of every order $t \geq 11$ which satisfies only the identity $x^2 = x$ belonging to E. The cases n = 6, 7, 8, 9, and 10 are handled easily by example.

<u><(12)>, <(13)>, <(23)></u>. As we have already seen, the identities $xy = yx$, $(yx)x = y$, and $x(xy) = y$ are conjugate and so it suffices to determine the spectrum for <(12)>. It is a trivial exercise to show that an idempotent commutative quasigroup of order n exists if and only if n is odd. Now there is only one idempotent commutative quasigroup of order 1 or 3 and both are totally symmetric and so we begin with 5. Let (Q,o) be the idempotent commutative quasigroup of order 5 given by the accompanying table.

(Q,o) =

o	1	2	3	4	5
1	1	4	2	5	3
2	4	2	5	3	1
3	2	5	3	1	4
4	5	3	1	4	2
5	3	1	4	2	5

Then (Q,o) satisfies $x^2 = x$ and $xy = yx$ only. Allan Cruse [4] has shown that an idempotent quasigroup of order n can be embedded in an idempotent commutative quasigroup of order t for <u>every odd</u> $t \geq 2n + 1$. Embedding (Q,o) in an idempotent commutative quasigroup of order t for every $t \geq 11$ gives an idempotent commutative quasigroup of every odd order $t \geq 11$ which satisfies $x^2 = x$ and $xy = yx$ only. The cases n = 7 and 9 can be handled by example.

<u><(123)></u>. Any quasigroup satisfying the identities $x^2 = x$ and $x(yx) = y$ (i.e., <u>any</u> idempotent semisymmetric quasigroup) is called a <u>Mendelsohn</u> quasigroup (after N. S. Mendelsohn [13]). If (Q,o) is a Mendelsohn quasigroup of order n, then the identity $x(yx) = y$ partitions $(Q \times Q) \setminus \{(i,i) | \text{all } i \in Q\}$ into 3-element sets of the form $\{(a,b),(b,c),(c,a)\}$ where $a \circ b = c$, $b \circ c = a$, and $c \circ a = b$. Hence $n \equiv 0$ or $1 \pmod 3$ is a necessary condition for the existence of a Mendelsohn quasigroup. A careful look at the construction (in section 3) of quasigroups satisfying <u>exactly</u> the semi-symmetric identity $x(yx) = y$ shows that in fact these quasigroups are idempotent in the cases where n $\equiv 0$ or $1 \pmod 3$. Hence this construction produces Mendelsohn quasigroups which are <u>not</u> totally symmetric for all $n \equiv 0$ or $1 \pmod 3$ and $n \geq 12$. It is not difficult to show that there <u>does not exist</u> a Mendelsohn quasigroup of order 6 and the cases n = 7, 9, and 10 can be handled by example. Finally, the quasigroup of order 4 in the semisymmetric construction in section 3 is <u>exactly</u> Mendelsohn <u>while</u> the idempotent quasigroups of orders 1 and 3 are totally symmetric.

$\underline{S_3}$. Any quasigroup satisfying the identities

$x^2 = x$ idempotent law

$$\left.\begin{aligned} x(xy) &= y \\ (yx)x &= y \end{aligned}\right\} \text{ Steiner identities}$$

is called a Steiner quasigroup. The following equivalence between Steiner quasigroups and Steiner triple systems (or simply triple systems) is very well known [8]. Let (S,o) be a Steiner quasigroup and define a collection of 3-element subsets t of S as follows: $\{x,y,z\} \in t$ if and only if $x \circ y = y \circ x = z$, $x \circ z = z \circ x = y$, and $y \circ z = z \circ y = x$. Then the collection of triples t has the property that every pair of distinct elements of S belongs to exactly one triple of t. Conversely, if (S,t) is a triple system and we define a binary operation $\circ$ on S by $a \circ a = a$ for all $a \in S$ and, if $a \neq b$, $a \circ b = c$ if and only if $\{a,b,c\} \in t$, then (S,o) is a Steiner quasigroup. For technical reasons it is easier to describe the construction of triple systems than it is to describe the construction of Steiner quasigroups and so we now irreversibly switch over. All of what follows is, of course, well-known (see [8] for example); but keep in mind that this is a survey paper.

It is a trivial matter to see that a necessary condition for the existence of a triple system (S,t) is that $|S| = n \equiv 1$ or $3 \pmod 6$ and $|t| = n(n-1)/6$. The following two constructions show sufficiency. They are slight modifications of constructions due to R. C. Bose [3] and Th. Skolem [14].

The Bose Construction. Let (Q,o) be an idempotent commutative quasigroup of order $2v + 1$ where $Q = \{1,2,3, \ldots, 2v+1\}$. Let $S = Q \times \{1,2,3\}$ and define a collection of triples t of S by:

(1) $\{(x,1), (x,2), (x,3)\} \in t$ for every $x \in Q$; and

(2) if $x \neq y$, the three triples $\{(x,1),(y,1),(x \circ y,2)\}$, $\{(x,2), (y,2), (x \circ y,3)\}$, and $\{(x,3), (y,3), (x \circ y,1)\} \in t$.

It is straightforward to see that (S,t) is a triple system of order $6v + 3$.

The Skolem Construction. A quasigroup (Q,o) is called half-idempotent (and we say that it satisfies the identity $(1/2)x^2 = x$) if and only if $|Q| = 2v$, $Q = \{1,2,\ldots,2v\}$, and

$$x \circ x = \begin{cases} x, & x \leq v, \text{ and} \\ x-v, & x > v. \end{cases}$$

It is quite easy to construct a half-idempotent commutative quasigroup of every order $2v$. Just rename the elements in the Cayley table of $(Z_{2v},+)$. Now let (Q,o) be a half-idempotent commutative quasigroup of order $2v$, set $S = \{\infty\} \cup (Q \times \{1,2,3\})$, and define a collection of triples t of S by:

(1) $\{(x,1), (x,2), (x,3)\} \in t$ for every $x \in Q$ and $x \leq v$;

(2) for each $x > v$, the three triples $\{\infty,(x,1),(x-v,2)\}$, $\{\infty,(x,2),(x-v,3)\}$, and $\{\infty,(x,3),(x-v,1)\} \in t$; and

(3) if $x \neq y$, the three triples $\{(x,1),(y,1),(x \circ y,2)\}$, $\{(x,2),(y,2),(x \circ y,3)\}$, and $\{(x,3),(y,3),(x \circ y,1)\} \in t$.

As with the Bose Construction there is no difficulty in showing that (S,t) is a triple system of order $6v + 1$.

Combining the Bose and Skolem Constructions gives a Steiner triple system (= Steiner quasigroup = idempotent and totally symmetric quasigroup) of every order $n \equiv 1$ or $3 \pmod 6$.

Remarks. One final comment on the results in sections 3 and 4. For each subgroup H of S_3 we are not constructing (idempotent) quasigroups which are at least invariant under conjugation by H but (idempotent) quasigroups whose conjugate invariant subgroup is precisely H. There is a lot of difference! For example, in section 3, if for each subgroup H of S_3 we are interested only in constructing quasigroups which are at least invariant under conjugation by H, we could always use totally symmetric quasigroups. The spectrum for each subgroup H would then be the positive integers and there would literally be no problem to look at! In section 4 the definition of a Mendelsohn quasigroup requires only that the quasigroup satisfy the identities $x^2 = x$ and $x(yx) = y$ and these quasigroups exist (as was pointed out) for all $n \equiv 0$ or $1 \pmod 3$ except $n = 6$, for which no such quasigroup exists. However, in order for a Mendelsohn quasigroup to have $\langle(123)\rangle$ as its conjugate invariant subgroup it cannot satisfy any of identities $xy = yx$, $x(xy) = y$, or $(yx)x = y$ (which is equivalent to saying it is not totally symmetric). And so, in section 4 we are constructing Mendelsohn quasigroups which are not Steiner quasigroups.

5. $n^2 \times 4$ orthogonal arrays and identities. An $n^2 \times 4$ orthogonal array is just an $n^2 \times 4$ array with the property that if we run our fingers down any two columns the n^2 ordered pairs we get are all distinct. From now on "orthogonal array" means $n^2 \times 4$ orthogonal array. Now it is well-known that an $n^2 \times 4$ orthogonal array is equivalent to a pair of orthogonal quasigroups. The equivalence is the following: Let A be an orthogonal array and label the columns in any order R, C, o_1, and o_2 and define a pair of quasigroups (Q,o_1) and (Q,o_2) as follows: For each ordered pair (x,y) find the unique row of A with x in column R and y in column C and define $x \, o_1 \, y = z$ and $x \, o_2 \, y = w$, where z is the entry in column o_1 and w is the entry in column o_2 in this row. The quasigroups (Q,o_1) and (Q,o_2) so constructed are orthogonal. Evidently, this construction can be reversed.

Example 5.1.

o_1	1	2	3	4
1	1	3	4	2
2	4	2	1	3
3	2	4	3	1
4	3	1	2	4

o_2	1	2	3	4
1	4	1	2	3
2	3	2	1	4
3	1	4	3	2
4	2	3	4	1

A =

R	C	o_1	o_2
1	1	1	4
1	2	3	1
1	3	4	2
1	4	2	3
2	1	4	3
2	2	2	2
2	3	1	1
2	4	3	4
3	1	2	1
3	2	4	4
3	3	3	3
3	4	1	2
4	1	3	2
4	2	1	3
4	3	2	4
4	4	4	1

Just as with $n^2 \times 3$ orthogonal arrays we can permute the columns of an $n^2 \times 4$ orthogonal array and the result remains an $n^2 \times 4$ orthogonal array. A careful look at the orthogonal array A in the above example reveals that A is invariant under conjugation by $\alpha = (123)$; i.e., $A(123) = A$. In fact, A is invariant under conjugation by precisely $\langle(123)\rangle$. Now, just as was the case with $n^2 \times 3$ orthogonal arrays, we can ask for the spectrum for each subgroup of S_4. Not too surprizingly, the solution is a good deal more difficult and interesting than the solution for subgroups of S_3. Never-the-less, the problem has been completely solved in a series of four papers [9], [10], [11], and [12], the results in [12] being the coup de grâce.

A bit of reflection reveals that if $n \geq 2$, and therefore $n \geq 3$, the conjugate invariant subgroup of an $n^2 \times 4$ orthogonal array cannot contain any transpositions. Hence the only subgroups of S_4 we need investigate are $\langle 1 \rangle$, the trivial subgroup; $C_2 = \langle (ij)(st) \rangle$, any cyclic subgroup generated by the product of two disjoint transpositions; $C_3 = \langle (ijk) \rangle$, any cyclic subgroup generated by a cycle of length 3; $C_4 = \langle (ijst) \rangle$, any cyclic subgroup generated by a cycle of length 4; K_4, the Klein 4-group; and A_4, the alternating group of degree 4. In what follows we will discuss the subgroups $\langle 1 \rangle$, $\langle (12)(34) \rangle$, $\langle (123) \rangle$, and $\langle (1234) \rangle$ (as well as K_4 and A_4, of course), the other cases being translations. We begin with the accompanying table giving some necessary and sufficient conditions for n to belong to the spectrum of the given subgroup of S_4.

Subgroups of S_4 which can possibly be conjugate invariant subgroups for an $n^2 \times 4$ orthogonal array	Necessary conditions for n to belong to the spectrum of given subgroup.	Sufficient conditions for n to belong to the spectrum of given subgroups.
$\langle 1 \rangle$	$n \neq 1, 2$, or 6	A pair of orthogonal latin squares whose upper left hand corners look like | 1 | 2 | and | 3 | 2 |
$\langle (12)(34) \rangle$	$n \neq 1, 2, 3$, or 6	A self-orthogonal quasigroup of order n which does not satisfy either of the identities $(xy)(yx) = x$ or $(xy)(yx) = y$.
$\langle (123) \rangle$	$n \equiv 0$ or $1 \pmod 3$ and $n \neq 1$ or 6	A resolvable Mendelsohn quasigroup of order n.
K_4	$n \equiv 0$ or $1 \pmod 4$ and $n \neq 1$ or 5	A quasigroup of order n which satisfies $(xy)(yx) = x$ but not $x(yx) = y$.
C_4	$n \equiv 0$ or $1 \pmod 4$ and $n \neq 1$	A quasigroup of order n satisfying $(xy)(yx) = y$.
A_4	$n \equiv 1$ or $4 \pmod{12}$ and $n \neq 1$	A quasigroup of order n satisfying $x(yx) = y$ and $(xy)(yx) = x$.

Table 5.2

The following observations will speed up the verifications of Table 5.2.

(1) 1, 2, and 6 cannot belong to the spectrum of any of the subgroups listed in Table 5.2 since the conjugate invariant subgroup of

1	1	1	1

is S_4 and there does not exist a $2^2 \times 4$ or $6^2 \times 4$ orthogonal array.

(2) if α is any 1 - 1 mapping from $\{1,2,3,4\}$ onto $\{R, C, o_1, o_2\}$ we will write

$$\begin{pmatrix} 1 & 2 & 3 & 4 \\ 1\alpha & 2\alpha & 3\alpha & 4\alpha \end{pmatrix}$$

to indicate that we have labeled the ith column of an $n^2 \times 4$ orthogonal array with $i\alpha$.

(3) An orthogonal array A is invariant under conjugation by (12)(34) if and only if with labeling

$$\begin{pmatrix} 1 & 2 & 3 & 4 \\ R & C & o_1 & o_2 \end{pmatrix}$$

the quasigroups (Q,o_1) and (Q,o_2) are transposes.

(4) An orthogonal array which is invariant under conjugation by at least K_4 is equivalent to a quasigroup satisfying the identity $(xy)(yx) = x$ (Schröder identity). Let A be an orthogonal array invariant under conjugation by K_4 and label A

$$\begin{pmatrix} 1 & 2 & 3 & 4 \\ R & C & o_1 & o_2 \end{pmatrix}.$$

Then for any $a, b \in Q$, $(a,b,c,d) \in A$ implies (b,a,d,c), (c,d,a,b), and $(d,c,b,a) \in A$ also. Hence $(a\ o_1\ b)\ o_1\ (b\ o_1\ a) = c\ o_1\ d = a$ and so (Q,o_1) satisfies $(xy)(yx) = x$. On the other hand if (Q,o) is a quasigroup satisfying $(xy)(yx) = x$ then (Q,o) is self-orthogonal (an easy exercise). If A is the orthogonal array consisting of all rows of the form $(a,b,a\ o\ b,\ b\ o\ a)$, then A is at least invariant under conjugation by K_4. For if $(a,b,a\ o\ b,\ b\ o\ a) \in A$, then
$(b,a,b\ o\ a,\ a\ o\ b)$, $(a\ o\ b,\ b\ o\ a, (a\ o\ b)\ o\ (b\ o\ a),\ (b\ o\ a)\ o\ (a\ o\ b)) =$
$(a\ o\ b,\ b\ o\ a,\ a,\ b)$, and
$(b\ o\ a,\ a\ o\ b, (b\ o\ a)\ o\ (a\ o\ b),\ (a\ o\ b)\ o\ (b\ o\ a))$
$= (b\ o\ a,\ a\ o\ b,\ b,\ a) \in A$ also.

(5) An orthogonal array which is invariant under conjugation by $C_4 = \langle(1234)\rangle$ is equivalent to a quasigroup satisfying the identity $(xy)(yx) = y$ (Stein's 3rd law). It is <u>worth noting</u> that as soon as an orthogonal array A is invariant under conjugation by C_4 then C_4 <u>is</u> the conjugate invariant subgroup of A. Now let A be invariant under conjugation by C_4 and label A

$$\begin{pmatrix} 1 & 2 & 3 & 4 \\ R & o_1 & C & o_2 \end{pmatrix}.$$

Then if $a,b, \in Q$, $(a,c,b,d) \in A$ implies (c,b,d,a), (b,d,a,c), and $(d,a,c,b) \in Q$ also. Hence $(a\ o_1\ b)\ o_1\ (b\ o_1\ a) = c\ o_1\ d = b$ and so (Q,o_1) satisfies $(xy)(yx) = y$. The other direction is handled as in (4) above.

We now plunge into a line by line verification of Table 5.2.

<u><1></u>. Let (Q,o_1) and (Q,o_2) be a pair of orthogonal quasigroups such that the upper left hand corners of their associated latin squares look like $\begin{array}{|c|c|}\hline 1 & 2 \\ \hline \end{array}$ and $\begin{array}{|c|c|}\hline 3 & 2 \\ \hline \end{array}$. If A is the orthogonal array with rows $(a,b,a\ o_1\ b,\ a\ o_2\ b)$ then $(1,1,1,3)$ and $(1,2,2,2)$ belong to A. It is now a trivial matter to see that A cannot be invariant under conjugation by any of $(12)(34)$, $(13)(24)$, $(14)(23)$, nor any cycle (ijk); which implies that A is invariant under (1) only.

<u>$\langle(12)(34)\rangle$</u>. According to comment (3), a necessary and sufficient condition for the existence of an $n^2 \times 4$ orthogonal array which is invariant under conjugation by <u>at least</u> $\langle(12)(34)\rangle$ is the existence of a self-orthogonal quasigroup of order n. Since no such quasigroup of order 3 exists, 3 <u>cannot</u> belong to the spectrum of $\langle(12)(34)\rangle$. Now let (Q,o) be a self-orthogonal quasigroup which does <u>not</u> satisfy either of the identities $(xy)(yx) = x$ or $(xy)(yx) = y$ and form the orthogonal array A consisting of the rows $(a,b,a\ o\ b,b\ o\ a)$. Then A is invariant under conjugation by <u>at least</u> $\langle(12)(34)\rangle$. The subgroups which contain $\langle(12)(34)\rangle$ properly are K_4, A_4, and $\langle(1324)\rangle$. However, A <u>cannot</u> be invariant under conjugation by K_4 since this would imply that (Q,o) satisfies $(xy)(yx) = x$ <u>and</u> A <u>cannot</u> be invariant under conjugation by $\langle(1324)\rangle$ since this would imply that

(Q,o) satsfies (xy)(yx) = y. Hence A has conjugate invariant subgroup <(12)(34)>.

<u><(123)></u>. Let A be an orthogonal array with conjugate invariant subgroup <(123)>. Label A by

$$\begin{pmatrix} 1 & 2 & 3 & 4 \\ R & C & o_1 & o_2 \end{pmatrix}.$$

Then (Q,o_1) is a Mendelsohn quasigroup. This is easy to see. Let (a,a,b,c) ε A. Then (a,b,a,c) ε A implies a = b and so (Q,o_1) is idempotent. Now suppose $a\ o_1\ b = c$. Then (a,b,c,d) ε A (for some d) implies (b,c,a,d) and (c,a,b,d) ε A. Hence $b\ o_1\ (a\ o_1\ b) = b\ o_1\ c = a$ and so (Q,o_1) satisfies x(yx) = y. Hence n ≡ 0 or 1 (mod 3), n ≠ 6 is a necessary condition for the existence of an $n^2 \times 4$ orthogonal array with conjugate invariant subgroup <(123)>.

Now if (Q,o) is a Mendelsohn quasigroup, as we have already seen in section 4, the identity x(yx) = y partitions $(Q \times Q) \setminus \{(i,i) \mid \text{all } i \in Q\}$ into sets of the form {(a,b),(b,c),(c,a)} where a o b = c, b o c = a, and c o a = b. Such sets will be called <u>directed triples</u>, and will be denoted by (a,b,c), (b,c,a), or (c,a,b). If |Q| = n ≡ 0(mod 3) a <u>parallel class</u> is a collection of n/3 directed triples (a,b,c) such that the n/3 sets {a,b,c} partition Q. If |Q| = n ≡ 1(mod 3) a <u>parallel class</u> is a collection of (n-1)/3 directed triples (a,b,c) such that the (n-1)/3 sets {a,b,c} partition $Q \setminus \{x\}$ for some x ε Q. The element x is called the <u>deficiency</u> of the parallel class. A simple counting argument shows that no two parallel classes can have the same deficiency. We will <u>need</u> this fact later. A Mendelsohn quasigroup is called <u>resolvable</u> provided the directed triples (a,b,c) can be partitioned into parallel classes. So let (Q,o) be a resolvable Mendelsohn quasigroup of order n. There are precisely n(n-1)/3 directed triples (a,b,c). If |Q| = n ≡ 0(mod 3) denote the parallel classes by $\pi_1, \pi_2, \ldots, \pi_{n-1}$ and define an orthogonal array A with rows $\{(a,b,a\ o\ b,x) \mid (a,b,a\ o\ b) \in \pi_x\} \cup \{(i,i,i,n) \mid \text{all } i \in Q\}$. If |Q| = n ≡ 1(mod 3) denote the parallel classes by $\pi_1, \pi_2, \ldots, \pi_n$ where

x is the deficiency in parallel class π_x (recall that no two parallel classes have the same deficiency). Let α be any permutation on Q such that $1\alpha = n$ and define an orthogonal array A with rows $\{(a,b,a \circ b, x\alpha) \mid (a,b,a \circ b) \in \pi_x\} \cup \{(i,i,i,i\alpha) \mid \text{all } i \in Q\}$. In either case A is an orthogonal array invariant under conjugation by $\langle(123)\rangle$. To see that $C_3 = \langle(123)\rangle$ is in fact the conjugate invariant subgroup of A it is only necessary to show that A is not invariant under conjugation by (12)(34), since the only admissible subgroup of S_4 which contains C_3 properly is A_4. But this is immediate since by construction $(1,1,1,n) \in A$ so that if A were invariant under conjugation by (12)(34) we would also have $(1,1,n,1) \in A$ which, of course, cannot be.

Remark. The only requirement here is that the Mendelsohn quasigroup be resolvable. It may or may not be totally symmetric.

K_4. Necessary conditions are easily obtained by the method for C_4 in [12] (a more technical argument is given in [11]). The generators of orbits of K_4 acting as a permutation group on A look like: $g_1 = (a,a,a,a)$, $g_2 = (a,a,b,b)$, $g_3 = (a,b,a,b)$, $g_4 = (a,b,b,a)$, and $g_5 = (a,b,c,d)$, giving orbits of lengths 1,2,2,2 and 4 respectively. So, if there are n_i orbits of each type, $1 \leq i \leq 5$, then

$$n^2 = n_1 + 2n_2 + 2n_3 + 2n_4 + 4n_5.$$

But, as there are n rows with the same 1st and j^{th} columns for $j = 2$, 3 or 4 it follows that

$$n = n_1 + 2n_2 = n_1 + 2n_3 = n_1 + 2n_4.$$

Hence, $n_2 = n_3 = n_4$ and so $n(n-1)/4 = n_4 + n_5$, showing that $n \equiv 0$ or $1 \pmod 4$.

Sufficiency follows almost immediately from comment (4). Suppose that $(Q,\circ)$ satisfies $(xy)(yx) = x$ but not $x(yx) = y$. Then the orthogonal A array with rows $(a,b,a \circ b, b \circ a)$ is invariant under conjugation by at least K_4. Since the only admissible subgroup of S_4 which contains K_4 properly is A_4 it is only necessary to show that A is not invariant under conjugation by (123). But this cannot be since this would imply that $(Q,\circ)$ satisfies $x(yx) = y$.

Remark. In the above argument, the fact that the orthogonal array A is invariant under conjugation by at least K_4 depends solely on the fact that the quasigroup (Q,o) satisfies the identity $(xy)(yx) = x$. Hence $n \equiv 0$ or $1 \pmod 4$ is a necessary condition for the existence of a quasigroup of order n satisfying $(xy)(yx) = x$. The requirement that (Q,o) not satisfy $x(yx) = y$ does not alter this necessary condition.

C_4. In this case, the generators of orbits of C_4 acting as a permutation group on A look like g_1, g_3 and g_5 above and $g_6 = (a,a,b,c)$, giving orbits of lengths 1,2,4, and 4 respectively. Counting as before, we find, in an obvious notation

$$\begin{cases} n^2 = n_1 + 2n_3 + 4n_5 + 4n_6, \text{ and} \\ n = n_1 + 2n_3. \end{cases}$$

Hence, $n(n-1)/4 = n_5 + n_6$ and, once again, $n \equiv 0$ or $1 \pmod 4$.

Here, sufficiency is a consequence of comment (5).

A_4. Let A be an $n^2 \times 4$ orthogonal array based on $Q = \{1,2,3,\ldots,n\}$ which is invariant under conjugation by A_4. If we label A by

$$\begin{pmatrix} 1 & 2 & 3 & 4 \\ R & C & o_1 & o_2 \end{pmatrix}$$

it is an easy exercise to see that (Q,o_1) satisfies the identities

$$\begin{cases} x(yx) = y, \text{ and} \\ (xy)(yx) = x\ . \end{cases}$$

It is well-known that a quasigroup satisfying these identities is equivalent to a block design with block size 4. Since such block designs exist if and only if $n \equiv 1$ or $4 \pmod{12}$ [6] necessity is established. On the other hand if (Q,o) is a quasigroup satisfying the above two identities and we define an orthogonal array A with rows $(a,b,a \circ b, b \circ a)$ it is no big deal to verify that A is invariant under conjugation by A_4.

6. The spectrum for subgroups of S_4. The accompanying table gives the spectrum for each admissible subgroup of S_4. In view of Table 5.2 we need only verify sufficiency; which is achieved by constructing quasigroups which satisfy the conditions given in Table 5.2. As might be expected, this is a vastly more complicated undertaking than in sections 3 and 4 and in many cases involves heavy doses of the "hammer and tongs" technique. (The "possible exceptions" for K_4 and C_4 means that, as yet, no one has constructed quasigroups of these orders satisfying the given sufficient conditions.) Since this is a survey paper, we omit the details (the interested reader can go straight to the original papers) and content ourselves with an example in each case.

subgroups of S_4	spectrum
$\langle 1 \rangle$	all $n \neq 1,2$, or 6 [12].
$\langle (12)(34) \rangle$	all $n \neq 1,2,3$, or 6 [12]
$\langle (123) \rangle$	all $n \equiv 0$ or $1 \pmod 3$ except 1 and 6 [1,2;12]
K_4	all $n \equiv 0$ or $1 \pmod 4$ except 1 and 5 and possibly 12 and 21 [11]
C_4	all $n \equiv 0$ or $1 \pmod 4$ except 1 and possibly 12 and 48 [12]
A_4	all $n \equiv 1$ or $4 \pmod{12}$ except 1 [6,10]

Table 6.1

<(1)>.

o_1	1	2	3
1	1	2	3
2	2	3	1
3	3	1	2

o_2	1	2	3
1	3	2	1
2	1	3	2
3	2	1	3

Pair of orthogonal quasigroups whose upper left hand corners look like | 1 | 2 | and | 3 | 2 |.

R	C	o_1	o_2
1	1	1	3
1	2	2	2
1	3	3	1
2	1	2	1
2	2	3	3
2	3	1	2
3	1	3	2
3	2	1	1
3	3	2	3

<(12)(34)>.

o	1	2	3	4
1	2	1	3	4
2	4	3	1	2
3	1	2	4	3
4	3	4	2	1

Self-orthogonal but does not satisfy either of the identities $(xy)(yx) = x$ or $(xy)(yx) = y$.

R	C	aob	boa
1	1	2	2
1	2	1	4
1	3	3	1
1	4	4	3
2	1	4	1
2	2	3	3
2	3	1	2
2	4	2	4
3	1	1	3
3	2	2	1
3	3	4	4
3	4	3	2
4	1	3	4
4	2	4	2
4	3	2	3
4	4	1	1

<(123)>

o	1	2	3
1	1	3	2
2	3	2	1
3	2	1	3

Resolvable Mendelsohn quasigroup

π_1: (1,2,3)
π_2: (1,3,2)

R	C	aob	
1	2	3	1
2	3	1	1
3	1	2	1
1	3	2	2
3	2	1	2
2	1	3	2
1	1	1	3
2	2	2	3
3	3	3	3

o	1	2	3	4
1	1	3	4	2
2	4	2	1	3
3	2	4	3	1
4	3	1	2	4

Resolvable Mendelsohn quasigroup
π_1: (2,4,3), deficiency 1
π_2: (1,3,4), deficiency 2
π_3: (1,4,2), deficiency 3
π_4: (1,2,3), deficiency 4.

$$\alpha = \begin{pmatrix} 1 & 2 & 3 & 4 \\ 4 & 2 & 3 & 1 \end{pmatrix}.$$

R	C	aob	
2	4	3	4
4	3	2	4
3	2	4	4
1	3	4	2
3	4	1	2
4	1	3	2
1	4	2	3
4	2	1	3
2	1	4	3
1	2	3	1
2	3	1	1
3	1	2	1
1	1	1	4
2	2	2	2
3	3	3	3
4	4	4	1

$\underline{K_4}$.

o	1	2	3	4
1	2	3	1	4
2	4	1	3	2
3	3	2	4	1
4	1	4	2	3

Satisfies the identity $(xy)(yx) = x$ but <u>not</u> $x(yx) = y$.

R	C	aob	boa
1	1	2	2
1	2	3	4
1	3	1	3
1	4	4	1
2	1	4	3
2	2	1	1
2	3	3	2
2	4	2	4
3	1	3	1
3	2	2	3
3	3	4	4
3	4	1	2
4	1	1	4
4	2	4	2
4	3	2	1
4	4	3	3

$\underline{C_4}$.

o	1	2	3	4
1	1	3	2	4
2	4	2	3	1
3	3	1	4	2
4	2	4	1	3

Satisfies the identity $(xy)(yx) = y$.

R	aob	C	boa
1	1	1	1
1	3	2	4
1	2	3	3
1	4	4	2
2	4	1	3
2	2	2	2
2	3	3	1
2	1	4	4
3	3	1	2
3	1	2	3
3	4	3	4
3	2	4	1
4	2	1	4
4	4	2	1
4	1	3	2
4	3	4	3

$\underline{A_4}$.

o	1	2	3	4
1	1	3	4	2
2	4	2	1	3
3	2	4	3	1
4	3	1	2	4

Satisfies
$x(yx) = y$, and
$(xy)(yx) = x$.

R	C	aob	boa
1	1	1	1
1	2	3	4
1	3	4	2
1	4	2	3
2	1	4	3
2	2	2	2
2	3	1	4
2	4	3	1
3	1	2	4
3	2	4	1
3	3	3	3
3	4	1	2
4	1	3	2
4	2	1	3
4	3	2	1
4	4	4	4

7. The identities $(xy)(yx) = x$ and $(xy)(yx) = y$. In the accepted parlance, the spectrum of the quasigroup identity $u(x,y) = v(x,y)$ is the set of all n such that there exists a quasigroup of order n satisfying at least the identity $u(x,y) = v(x,y)$. People in universal algebra (among others) love to determine the spectrum of interesting identities; and two identities which have recently attracted quite a bit of attention are $(xy)(yx) = x$ and $(xy)(yx) = y$. The results (and related work) in this set of notes handle both identities. As we have already noted in section 5, $n \equiv 0$ or $1 \pmod 4$ is necessary for the existence of a quasigroup of order n satisfying $(xy)(yx) = x$. Since no such quasigroup of order 5 exists [11] and the quasigroup of order 1 satisfies everything, it follows (Tables 5.2 and 6.1) that the spectrum for $(xy)(yx) = x$ is all $n \equiv 0$ or $1 \pmod 4$ except 5 and possibly 12 and 21. Since a positive integer n belongs to the spectrum of C_4 if and only if there exists a quasigroup of order n satisfying $(xy)(yx) = y$ and (as we have already noted) the quasigroup of order 1 satisfies everything, the spectrum for $(xy)(yx) = y$ is all $n \equiv 0$ or $1 \pmod 4$ except possibly 12 and 48. In keeping with this set of notes we tabulate these results in the following table.

identity	spectrum
$(xy)(yx) = x$	all $n \equiv 0$ or 1 (mod 4) except 5 and possibly 12 and 21 [11].
$(xy)(yx) = y$	all $n \equiv 0$ or 1 (mod 4) except possibly 12 and 48. [12].

Table 7.1

8. Concluding remarks. The author frankly admits that this set of notes lacks detail. This is by design! The author is attempting

to present the flavor of the topic and not a tedious exercise in mental gymnastics. Therefore, it is not necessary for anyone to have a nervous breakdown because of lack of detail, lack of historical perspective, or somebody's name is missing. The reader interested in pursuing the subject will find plenty of details in the references, including references not listed in this set of notes. So be it. Finally, a lot of work has been done on the problem of determining the spectrum for the admissible subgroups of S_5. However, since the complete solution is still in flux and the author has already used up around 30 pages it is time to stop.

Acknowledgement. Research supported by NSF Grant MCS 80-03053.

References.

[1] F.E. Bennett and D. Sotteau, Almost resolvable decompositions of K_n^*, J. Combinatorial Theory Ser. B 30(1981), 228-232.

[2] J.C. Bermond, A. Germa, and D. Sotteau, Resolvable decompositions of K_n^*, J. Combinatorial Theory Ser. A 26(1979), 179-185.

[3] R.C. Bose, On the construction of balanced incomplete block designs, Ann. Eugenics 9(1939), 353-399.

[4] Allan Cruse, On embedding incomplete symmetric latin squares, J. Combinatorial Theory Ser. A 16(1974), 18-27.

[5] Trevor Evans, Embedding incomplete latin squares, Amer. Math. Monthly 67(1960), 958-961.

[6] H. Hanani, The existence and construction of balanced incomplete block designs, Ann. Math. Stat. 32(1961), 361-386.

[7] A.J.W. Hilton, Embedding an incomplete diagonal latin square in a complete diagonal latin square, J. Combinatorial Theory Ser. A 15(1973), 121-128.

[8] C.C. Lindner, A survey of embedding theorems for Steiner systems, Annals of Discrete Mathematics 7(1980), 175-202.

[9] C.C. Lindner and D. Steedley, On the number of conjugates of a quasigroup, Algebra Universalis 5(1975), 191-196.

[10] C.C. Lindner and E. Mendelsohn, On the conjugates of an $n^2 \times 4$ orthogonal array, Discrete Math. 20(1977), 123-132.

[11] C.C. Lindner, N.S. Mendelsohn, and S.R. Sun, On the construction of Schroeder quasigroups, Discrete Math. 32(1980), 271-280.

[12] C.C. Lindner, R.C. Mullin, and D.G. Hoffman, The spectra for the conjugate invariant subgroups of $n^2 \times 4$ orthogonal arrays, Canad. J. Math. 32(1980), 1126-1139.

[13] N.S. Mendelsohn, A natural generalization of Steiner triple systems, Computers in Number Theory, Academic Press, New York, 1971, 323-338.

[14] Th. Skolem, Some remarks on the triple systems of Steiner, Math. Scand. 6(1958), 273-280.

BOUNDS ON PERMANENTS, AND THE NUMBER OF 1-FACTORS AND 1-FACTORIZATIONS OF BIPARTITE GRAPHS

A. Schrijver

Instituut voor Actuariaat en Econometrie,
Universiteit van Amsterdam,
Jodenbreestraat 23,
1011 NH Amsterdam, Holland.

Abstract. We give a survey of some recent developments on bounds for permanents (Falikman-Egorychev, Voorhoeve, Bang, Brègman), and show some related results on counting 1-factors (perfect matchings), 1-factorizations (edge-colourings), and eulerian orientations of graphs.

1. INTRODUCTION.

The *permanent* of a square matrix $A = (a_{ij})_{i,j=1}^{n}$ is given by:

(1) $$\text{per}A = \sum_{\sigma \in S_n} \Pi_{i=1}^{n} a_{i\sigma(i)},$$

where S_n denotes the collection of all permutations of $\{1,\ldots,n\}$.

Despite its appearance as the simpler twin-brother of the determinant function, the permanent turns out to be much less tractable. Whereas a determinant can be calculated quickly (in polynomial time, with Gaussian elimination), determining the permanent is difficult ("number-P-complete" - see Valiant [27]). As yet, its algebraic behaviour appeared to a large extent unmanageable, and its algebraic relevance moderate.

Most interest in permanents came from the famous Van der Waerden conjecture on the minimum permanent of doubly stochastic matrices (see below). This conjecture was unsolved for more than fifty years, which, as contrasted with its simple form, also contributed to the image of intractability of permanents. Recently, Falikman and Egorychev were able to prove this conjecture, using a classical inequality of Alexandroff and Fenchel. The proof with eigenvalue techniques also revealed some unexpected nice algebraic behaviour of the permanent function.

In fact, lower and upper bounds form a field where a large part of the successes in controlling permanents have been obtained, also by the work of, e.g., Bang, Brègman and Voorhoeve. In this paper we discuss some of the bounds for the permanent function, and for the related numbers of 1-factors and 1-factorizations of bipartite graphs. Especially, we survey some recent work in this field.

The book by Minc [21] gives an excellent survey of what is known on permanents until 1978. Van Lint [15] gave a survey of bounds on permanents known in 1974. For some more historical remarks, see Van Lint [17].

In this introduction we first give a brief survey.

Van der Waerden's conjecture. In 1926 Van der Waerden [30] posed the following conjecture: if A is a doubly stochastic matrix of order n, then

$$\text{(2)} \qquad \mathrm{per}A \geq n!/n^n,$$

and equality only holds for $A = \frac{1}{n}J$ (J being the all-one matrix). A matrix is *doubly stochastic* if it is nonnegative and all row and column sums are 1.

As the permanent function is not convex, the Kuhn-Tucker theory (Lagrange multipliers) yields only necessary conditions for the doubly stochastic matrices minimizing the permanent. The conjecture raised a stream of research, especially during the last twenty years. In 1978, as a prelude, the lower bound of e^{-n} was proved by Bang [2] and Friedland [10], which bound is asymptotically equal to Van der Waerden's conjectured lower bound $n!/n^n$, by Stirling's formula. Ultimately in 1979 and 1980, Falikman [8] and Egorychev [6] published proofs of Van der Waerden's conjecture.

The basis for their proofs is a permanent inequality, which is a special case of an inequality for "mixed volumes" of convex bodies, found in the thirties by Fenchel [9] and Alexandroff [1] (cf. Busemann [5]). Let B be an $n\times(n-2)$-matrix, and let x and y be column vectors of length n. If B and x are nonnegative, then

$$\text{(3)} \qquad \mathrm{per}^2(B,x,y) \geq \mathrm{per}(B,x,x).\mathrm{per}(B,y,y).$$

(This can be seen to be equivalent to: the function $x \longrightarrow \sqrt{\mathrm{per}(B,x,x)}$ is concave on the nonnegative orthant.) The inequality (3) can be proved directly with an interesting eigenvalue technique ([12],[16]).

On the other hand, Marcus and Newman [19] and London [18] had shown that if A is a *minimizing* matrix (i.e. a doubly stochastic matrix minimizing the permanent), then $\mathrm{per}A_{ij} \geq \mathrm{per}A$ for each $(n-1)\times(n-1)$-minor A_{ij} of A. Hence, if (B,x,y) is a minimizing matrix, then $\mathrm{per}(B,x,y) \leq \mathrm{per}(B,x,x)$ and $\mathrm{per}(B,x,y) \leq \mathrm{per}(B,y,y)$ (as we can expand these permanents by the last columns, just like determinants, but without sign problems). Therefore, by (3), we have $\mathrm{per}(B,x,y) = \mathrm{per}(B,x,x) = \mathrm{per}(B,y,y)$. This implies

(4) $$\mathrm{per}(B,\tfrac{1}{2}x+\tfrac{1}{2}y,\tfrac{1}{2}x+\tfrac{1}{2}y) = \tfrac{1}{4}\mathrm{per}(B,x,x)+\tfrac{1}{2}\mathrm{per}(B,x,y)+\tfrac{1}{4}\mathrm{per}(B,y,y) = \mathrm{per}(B,x,y)$$

(using the fact that the permanent is linear in the columns). Since the matrix $(B,\frac{1}{2}x+\frac{1}{2}y,\frac{1}{2}x+\frac{1}{2}y)$ is doubly stochastic, by (4) it is minimizing again. If we assume that we have chosen the matrix (B,x,y) so that the sum of its squared components (i.e., $\mathrm{Tr}((B,x,y)^T(B,x,y))$) is as small as possible, it follows that $x = y$ (as $\mathrm{Tr}((B,\frac{1}{2}x+\frac{1}{2}y,\frac{1}{2}x+\frac{1}{2}y)^T(B,\frac{1}{2}x+\frac{1}{2}y,\frac{1}{2}x+\frac{1}{2}y)) \leq \mathrm{Tr}((B,x,y)^T(B,x,y))$ with equality iff $x = y$). As the columns x and y were chosen arbitrarily, we know that all columns of (B,x,y) are equal, that is, it is $\frac{1}{n}J$.

By extending these methods Egorychev proved that $\frac{1}{n}J$ is the only minimizing matrix. In Section 2 we describe a complete proof of Van der Waerden's conjecture, where we have benefitted by the presentations of Knuth [12] and Van Lint [16,17].

Permanents combinatorially. The permanent can be put in a more combinatorial context as follows. For natural numbers k and n, denote

(5) Λ_n^k = the set of all nonnegative integral $n\times n$-matrices with all line sums equal to k

(*lines* are rows and columns). Then Falikman and Egorychev's lower bound is equivalent to:

(6) if $A \in \Lambda_n^k$ then $\mathrm{per}A \geq (\frac{k}{n})^n n!$.

Indeed, if $A\in\Lambda_n^k$, $\frac{1}{k}A$ is doubly stochastic, and hence $\mathrm{per}A = k^n\mathrm{per}(\frac{1}{k}A) \geq k^n n!/n^n$. Conversely, any rational doubly stochastic matrix of order n is equal to $\frac{1}{k}A$ for some k and some $A\in\Lambda_n^k$. Then (6) gives $\mathrm{per}(\frac{1}{k}A) = k^{-n}\mathrm{per}A \geq n!/n^n$. So $n!/n^n$ is a lower bound for rational doubly stochastic matrices, and hence, by continuity, it is a lower bound for all doubly stochastic matrices.

To obtain a more combinatorial interpretation, if A is in Λ_n^k, we can construct the bipartite graph G with vertices $v_1,\ldots,v_n,w_1,\ldots,w_n$, connecting v_i and w_j by a_{ij} (possibly parallel) edges. Then G is k-regular, and the permanent of A is equal to the number of perfect matchings in G.

In 1968, Erdös and Rényi [7] published the following conjecture, weaker than Van der Waerden's conjecture:

(7) there is an $\varepsilon > 0$ such that if $A \in \Lambda_n^k$ with $k \geq 3$, then $\operatorname{per}A \geq (1+\varepsilon)^n$.

This conjecture is implied by Van der Waerden's conjecture through (6), as $(k/n)^n n! \geq (k/e)^n$ by Stirling's formula.

The Erdös-Rényi conjecture was proved in 1978 independently by Voorhoeve [29] and by Bang [2] and Friedland [10]. As mentioned before, Bang and Friedland showed that $\operatorname{per}A \geq e^{-n}$ for each doubly stochastic matrix A of order n, and hence $\operatorname{per}A \geq (k/e)^n$ for each $A \in \Lambda_n^k$. This shows (7).

Voorhoeve showed:

(8) if $A \in \Lambda_n^3$ then $\operatorname{per}A \geq (\frac{4}{3})^n$.

In other words, any 3-regular bipartite graph with 2n vertices has at least $(4/3)^n$ perfect matchings. Or: if A is a doubly stochastic matrix of order n, with all components a multiple of 1/3, then $\operatorname{per}A \geq (4/9)^n$. Asymptotically, for $n \to \infty$, this is better than Falikman and Egorychev's and Bang and Friedland's lower bounds ($(3/e)^n$). The best lower bound for permanents of matrices in Λ_n^3 found before Voorhoeve's result was 3n-2 (Hartfiel and Crosby [11]). With König's theorem (see Remark 1 below) (8) implies that $\operatorname{per}A \geq (4/3)^n$ for all $A \in \Lambda_n^k$, $k \geq 3$, and hence the Erdös-Rényi conjecture follows.

In [26] it has been shown that the ground number 4/3 in (8) is best possible. More generally, let f(k) be the highest possible number such that $\operatorname{per}A \geq f(k)^n$ for all $A \in \Lambda_n^k$. Then

(9) $$f(k) \leq \frac{(k-1)^{k-1}}{k^{k-2}}.$$

Note that by Bang's result, $f(k) \geq k/e$, and by Voorhoeve's result, $f(3) \geq 4/3$. The latter bound combined with (9) gives $f(3) = 4/3$. Trivially we have $f(1) = f(2) = 1$. It is conjectured in [26] that equality holds in (9) for every k. That is:

(10) (Conjecture) if $A \in \Lambda_n^k$ then $\operatorname{per} A \geq \left(\frac{(k-1)^{k-1}}{k^{k-2}}\right)^n$.

This conjecture would give a bound asymptotic for k fixed and $n \to \infty$, while Falikman and Egorychev's lower bound, in the form (6), is asymptotic for n fixed, $k \to \infty$. Conjecture (10) implies a better lower bound for permanents of doubly stochastic matrices with all components being a multiple of 1/k.

Voorhoeve's method consists of a clever induction trick, which it is tempting to generalize to values of $k \geq 4$. However, in this direction no significant progress has been made as yet.

For a more extensive discussion of Voorhoeve's result and best lower bounds, see Section 3.

Bang's method and edge-colourings. The method of Bang [2] gives rise to some further graph-theoretic considerations.

Suppose you have given the first lesson of a course on graph theory. You have explained Euler's result on the existence of eulerian orientations, and you have given the definitions of regular and bipartite graphs, and of perfect matchings. Now as homework you ask: show that each 64-regular bipartite graph has a perfect matching. Is this a reasonable question for your students, whom you do not expect to discover for themselves the König-Hall theorem?

Yes, it is. They know that the 64-regular bipartite graph has an eulerian orientation. By deleting the edges oriented from the "red" points to the "blue" points, and by forgetting the orientation of the other edges, we are left with a 32-regular bipartite graph. By the same reasoning this 32-regular graph has a 16-regular spanning subgraph. And so on, until we have a 1-regular spanning subgraph, which is a perfect matching.

This idea can be extended from the *existence* of perfect matchings to *counting* perfect matchings, and also to counting 1-factorizations of regular bipartite graphs ([24]). This last can be seen as the graph-theoretic interpretation of the ideas, in matrix language, of Bang, which have led to his lower bound e^{-n}.

It also leads to the following. In [24] it is conjectured that if G is a k-regular bipartite graph with 2n points, then

(11) (Conjecture) G has at least $(k!^2/k^k)^n$ 1-factorizations.

By an averaging argument it can be shown that the ground number in (11), as a function of k, cannot be higher. Moreover, using the ideas described above, and using Voorhoeve's lower bound, it can be shown that (11) is true if k has no other prime factors than 2 and 3.

These results are described more extensively in Section 4.

<u>Brègman's upper bound</u>. Now we turn to upper bounds. It is easy to see that the maximum permanent over the doubly stochastic matrices is 1, which is attained, exclusively, by the permutation matrices. Similarly, the maximum permanent over matrices in Λ_n^k is equal to k^n.

The problem becomes more difficult if we go over to a further discretization. In 1963, Minc [20] posed as a conjecture:

(12) if A is a square (0,1)-matrix of order n, with row sums $r_1,\ldots,r_n$, then
$$\text{per}A \leq \Pi_{i=1}^n \, r_i!^{1/r_i}.$$

In 1973, Brègman [4] found a proof for this conjecture, using ideas from convex programming, and some theory of doubly stochastic matrices. In [23] a shorter proof was given, using elementary counting and the convexity of the function xlogx.

Note that (12) implies that

(13) if $A \in \Lambda_n^k$ and A is (0,1), then $\text{per}A \leq \left(k!^{1/k}\right)^n$.

The ground number here can be easily seen to be asymptotically best possible (for fixed k).

The proof of Brègman's upper bound is given in Section 5.

<u>Eulerian orientations</u>. Finally, as a further illustration of the methods, we consider bounds for the number of eulerian orientations of undirected graphs. Let G = (V,E) be a loopless, 2k-regular undirected graph, with |V| = n and |E| = m. Let $\varepsilon(G)$ denote the number of eulerian orientations of G. Let B be the n×m-incidence matrix of G, and let A be the m×m-matrix obtained from B by repeating each row k times. Then one easily sees:

(14) $$\varepsilon(G) = \frac{\mathrm{per}A}{k!^n}.$$

Now it can be shown that

(15) $$\left(2^{-k}\binom{2k}{k}\right)^n \le \varepsilon(G) \le \sqrt{\binom{2k}{k}}^{\,n}.$$

The upper bound can be derived straightforwardly from Brègman's bound (12) using (14). The lower bound in (15) is better than the one derived with (14) from the conjectured lower bound (10).

It can be shown moreover that the ground numbers in (15) are best possible. These results are described further in Section 6.

Throughout this paper, n denotes the order of the matrix in question. Furthermore, if the matrix A is given, A_{ij} denotes the minor of A obtained by deleting the i-th row and the j-th column of A.

REMARK 1. We here remark the following well-known facts.

(16) *Doubly stochastic matrices minimizing the permanent exist.*

This follows of course from the compactness of the set of doubly stochastic matrices, and from the continuity of the permanent function.

(17) *Each doubly stochastic matrix is a convex linear combination of permutation matrices.*

This result of Birkhoff [3] and Von Neumann [22] can be seen by induction on n. It suffices to show that each vertex of the polytope of doubly stochastic matrices is a convex linear combination of permutation matrices (and hence is a permutation matrix itself). Let $A = (a_{ij})_{i,j=1}^n$ be a vertex of this polytope. Then n^2 linearly independent inequalities in the system: $a_{ij} \ge 0$ $(i,j=1,\ldots,n)$, $\sum_i a_{ij} = 1$ $(j=1,\ldots,n)$, $\sum_j a_{ij} = 1$ $(i=1,\ldots,n)$, are satisfied with equality. So A has at least n^2-2n+1 zeros, and hence at least one row has n-1 zeros. So $a_{ij} = 1$ for some i,j. Then A_{ij} is doubly stochastic again, and by the induction hypothesis, it is a convex linear combination of permutation matrices of order n-1. Therefore, A itself is a convex linear combination of permutation matrices of order n.

(17) implies:

(18) perA > 0 *if* A *is a doubly stochastic matrix;* perA ≥ 1 *if* A *is in* Λ_n^k.

The second assertion is equivalent to a result of König [14]: each k-regular bipartite graph has a perfect matching. So for each $A \in \Lambda_n^k$ there exists an $A' \in \Lambda_n^{k-1}$ with $A' \le A$ (≤ component-wise). Inductively it implies that each k-regular bipartite graph has a k-edge colouring, which is another theorem of König [13].

2. FALIKMAN AND EGORYCHEV'S PROOF OF THE VAN DER WAERDEN CONJECTURE.

Van der Waerden's conjecture (2) was proved by Falikman [8] and Egorychev [6] (cf. Knuth [12] and Van Lint [16,17]). The ingredients are two results, the first one being a special case of an inequality for "mixed volumes" of convex bodies, due to Fenchel [9] and Alexandroff [1] (cf. Busemann [5]).

<u>THEOREM 1</u> (Alexandroff-Fenchel permanent inequality). *If* B *is a nonnegative* $n\times(n-2)$*-matrix,* x *and* y *are column vectors of length* n, *and* $x \ge 0$, *then*

(19) $$\mathrm{per}^2(B,x,y) \ge \mathrm{per}(B,x,x).\mathrm{per}(B,y,y).$$

If B *and* x *are strictly positive, equality holds in* (19) *if and only if* $y = \lambda x$ *for some* λ.

<u>PROOF</u>. The proof is by induction on n, the case n=2 being easy. Suppose the theorem has been proved for n-1. To prove (19), by continuity we may assume that all components of B and x are positive. Define the matrix $Q = (q_{ij})_{i,j=1}^n$ by:

(20) $$q_{ij} = \mathrm{per}(B,e_i,e_j),$$

where e_i and e_j denote the i-th and the j-th column standard basis vectors. So $\mathrm{per}(B,x,y) = x^T By$.

I. We first show that Q is nonsingular with exactly one positive eigenvalue (i.e., it defines a "Lorentz space"). To see that Q is nonsingular, assume that Qc = 0, that is:

(21) $\quad \mathrm{per}(B,c,e_j) = 0$

for $j = 1,\ldots,n$. Let $B = (C,z)$, where z is the last column of B (so C is an $n\times(n-3)$-matrix). Then for each $j = 1,\ldots,n$:

(22) $\quad 0 = \mathrm{per}^2(C,z,c,e_j) \geq \mathrm{per}(C,z,z,e_j).\mathrm{per}(C,c,c,e_j).$

The equality here follows from (21), and the inequality from our induction hypothesis: as e_j is the j-th standard basis vector, the matrices in (22) can be replaced by their (j,n)-th minors.

Since $\mathrm{per}(C,z,z,e_j) > 0$ (as C and z are positive), (22) gives that $\mathrm{per}(C,c,c,e_j) \leq 0$. As from (21) $\mathrm{per}(C,z,c,e_j) = 0$ for all j, we know:

(23) $\quad 0 = \mathrm{per}(C,c,c,z) = \sum_{j=1}^{n} z_j\mathrm{per}(C,c,c,e_j) \leq 0.$

As z is positive, (23) implies that $\mathrm{per}(C,c,c,e_j) = 0$ for all j. Hence the inequality in (22) holds with equality, for all j, and therefore, from the induction hypothesis, $c = \lambda z$ for some λ. If $\lambda \neq 0$ then $0 = \mathrm{per}(B,c,e_j) = \lambda\mathrm{per}(B,z,e_j) \neq 0$ (as B and z are positive), which is a contradiction. So $\lambda = 0$ and hence $c = 0$. Concluding $Qc = 0$ implies $c = 0$, and so Q is nonsingular.

Now, for each real number μ, let the matrix Q_μ be defined by:

(24) $\quad Q_\mu = (\mathrm{per}(\mu B+(1-\mu)J,e_i,e_j))_{i,j=1}^{n}$

(here J denotes the all-one $n\times(n-2)$-matrix). So $Q_1 = Q$. Since $\mu B+(1-\mu)J$ is a positive matrix for $0 \leq \mu \leq 1$, we know by the above that Q_μ is nonsingular for $0 \leq \mu \leq 1$. For $\mu = 0$, Q_μ is a matrix with zero diagonal and with all off-diagonal components equal to $(n-2)!$, and so it has exactly one positive eigenvalue. Therefore, as the shift of the spectrum of Q_μ is continuous in μ, also for $\mu = 1$ the matrix $Q_\mu = Q$ has exactly one positive eigenvalue.

II. We now prove the theorem. The inequality (19) is equivalent to:

(25) $\quad (x^TQy)^2 \geq (x^TQx).(y^TQy).$

This inequality holds trivially with equality if x and y are linearly dependent. If x and y are linearly independent, the (2-dimensional) linear hull of x and y intersects the ((n-1)-dimensional) linear hull of the eigenvectors of Q with negative eigenvalue in a nonzero vector (as Q has n-1 negative eigenvalues). Therefore

$$(26) \qquad (\lambda x+\mu y)^T Q(\lambda x+\mu y) < 0$$

for some λ,μ not both zero. Since $x^T Qx = \mathrm{per}(B,x,x) > 0$ (as $x > 0$), we know that $\mu \neq 0$. We may assume $\mu = 1$. Then the left hand side of (26) becomes a quadratic polynomial in λ, with positive main coefficient $x^T Qx$, and at least one negative value. Hence its discriminant is positive, which means that (25) holds, with strict inequality. □

A second ingredient for the proof of Van der Waerden's conjecture is a theorem due to Marcus and Newman [19] and London [18].

<u>THEOREM 2</u>. *If* A *is a doubly stochastic matrix minimizing the permanent, then* $\mathrm{per}A_{ij} \geq \mathrm{per}A$ *for each minor* A_{ij} *of* A.

<u>PROOF</u>. Let A be a minimizing matrix of order n. Consider the directed bipartite graph G with vertices $u_1,\ldots,u_n,v_1,\ldots,v_n$, and with arcs:

(27) (i) (u_i,v_j) iff $\mathrm{per}A_{ij} \leq \mathrm{per}A$;
(ii) (v_j,u_i) iff $a_{ij} > 0$ and $\mathrm{per}A_{ij} \geq \mathrm{per}A$.

Assume that, say, $\mathrm{per}A_{11} < \mathrm{per}A$. We first show that then the arc (u_1,v_1) of G is not contained in any directed cycle of G. For suppose that C is such a cycle. Let $\varepsilon > 0$, and

(28) (i) replace a_{ij} by $a_{ij}+\varepsilon$ if (u_i,v_j) belongs to C,
(ii) replace a_{ij} by $a_{ij}-\varepsilon$ if (v_j,u_i) belongs to C.

Let A_ε be the matrix arising in this way. Now $\mathrm{per}A_\varepsilon$ is a polynomial in ε, and:

$$(29) \qquad \mathrm{per}A_\varepsilon = \mathrm{per}A + \varepsilon\Big(\sum_{(u_i,v_j)\in C}\mathrm{per}A_{ij} - \sum_{(v_j,u_i)\in C}\mathrm{per}A_{ij}\Big) + O(\varepsilon^2) \qquad (\varepsilon\downarrow 0)$$

The coefficient of ε in (29) is negative, by (27) and as $\mathrm{per}A_{11} < \mathrm{per}A$ (the first summation is strictly smaller than $\frac{1}{2}|C|\mathrm{per}A$, and the second summation is at least $\frac{1}{2}|C|\text{per}A$). So by choosing ε small enough A_ε is doubly stochastic with $\text{per}A_\varepsilon < \text{per}A$, contradicting that A is minimizing.

So the arc (u_1,v_1) is not contained in any directed cycle. Let, say, $v_1,\dots,v_k,u_{t+1},\dots,u_n$ be the points of G which can be reached by a directed path from v_1. So $k,t \geq 1$, and G has no arcs (u_i,v_j) with $i \geq t+1$ and $j \geq k+1$, nor arcs (v_j,u_i) with $j \leq k$ and $i \leq t$. That is:

(30) (i) if $i \geq t+1$ and $j \geq k+1$ then $\text{per}A_{ij} > \text{per}A$;
(ii) if $i \leq t$ and $j \leq k$ then $a_{ij} = 0$ or $\text{per}A_{ij} < \text{per}A$.

Now:

(31)
$$(n-k-t)\text{per}A = \sum_{i>t}\sum_j a_{ij}\text{per}A_{ij} - \sum_{j\leq k}\sum_i a_{ij}\text{per}A_{ij} =$$
$$=\Big(\sum_{i>t}\sum_{j>k}a_{ij}\text{per}A_{ij}-\sum_{i\leq t}\sum_{j\leq k}a_{ij}\text{per}A_{ij}\Big)\geq\Big(\sum_{i>t}\sum_{j>k}a_{ij}-\sum_{i\leq t}\sum_{j\leq k}a_{ij}\Big)\text{per}A=$$
$$= \Big(\sum_{i>t}\sum_j a_{ij} - \sum_{j\leq k}\sum_i a_{ij}\Big)\text{per}A = (n-k-t)\text{per}A.$$

Here the inequality follows from (30). The equalities follow from $\sum_j a_{ij}=1$ and $\sum_j a_{ij}\text{per}A_{ij} = \text{per}A$ for all i (and similarly for j), and by crossing out equal terms in the summations.

Since the first and the last term in (31) are equal, the inequality is an equality. Hence, by (30), $a_{ij} = 0$ if $i \leq t$, $j \leq k$ or if $i > t$, $j > k$. Therefore, all terms in (31) are zero, and hence $n = k+t$.

Since $k,t \geq 1$ and $n = k+t$, it follows that $k,t \leq n-1$. Hence from (30), $\text{per}A_{nn} > \text{per}A > 0$. So there is a permutation σ of $\{1,\dots,n-1\}$ with $a_{i\sigma(i)} > 0$ for $i = 1,\dots,n-1$. As $k > (n-t)-1$ this implies that $a_{ij} > 0$ for at least one pair of $i \leq t$, $j \leq k$, contradicting what we showed above. □

(Alternatively, Theorem 2 can be proved using Kuhn-Tucker theory.)

Combining Theorems 1 and 2 gives the theorem of Falikman and Egorychev.

<u>THEOREM 3</u> (Falikman-Egorychev theorem). *If* A *is a doubly stochastic matrix of order* n, *then* $\text{per}A \geq n!/n^n$.

<u>PROOF</u>. We first show that if $A = (B,x,y)$ is a doubly stochastic matrix minimizing the permanent (where x and y are the last two columns of A), then

(32) $$\text{per}(B,x,y) = \text{per}(B,x,x) = \text{per}(B,y,y).$$

Indeed, by Theorem 2,

$$(33)\qquad \mathrm{per}(B,x,x) = \sum_i x_i \mathrm{per}(B,x,e_i) \geq \mathrm{per}(B,x,y)\sum_i x_i = \mathrm{per}(B,x,y).$$

Similarly, $\mathrm{per}(B,y,y) \geq \mathrm{per}(B,x,y)$. On the other hand, by Theorem 1, $\mathrm{per}^2(B,x,y) \geq \mathrm{per}(B,x,x)\mathrm{per}(B,y,y)$. Since $\mathrm{per}(B,x,y) > 0$ (cf. (18)), it follows that $\mathrm{per}(B,x,y) = \mathrm{per}(B,x,x) = \mathrm{per}(B,y,y)$.

(32) implies that:

$$(34)\qquad \mathrm{per}(B,\tfrac12x+\tfrac12y,\tfrac12x+\tfrac12y) = \tfrac14\mathrm{per}(B,x,x)+\tfrac12\mathrm{per}(B,x,y)+\tfrac14\mathrm{per}(B,y,y) = \mathrm{per}(B,x,y$$

Since $(B,\frac12x+\frac12y,\frac12x+\frac12y)$ is doubly stochastic again, it is again minimizing.

Now suppose we have chosen A such that $\sum_{i,j} a_{ij}^2 = \mathrm{Tr}A^TA$ is as small as possible (this is possible by compactness). Assume $A \neq (1/n)J$. Without loss of generality, $A = (B,x,y)$ with $x \neq y$. By the above, the matrix $A' :=$ $(B,\frac12x+\frac12y,\frac12x+\frac12y)$ is minimizing again. However, $\mathrm{Tr}(A'^TA') < \mathrm{Tr}(A^TA)$ (as $x \neq y$), contradicting our assumption.

Therefore, $A = (1/n)J$, and $\mathrm{per}A = n!/n^n$. □

Extension of these arguments gives the uniqueness of $(1/n)J$ as a minimizing matrix. Suppose there exists a doubly stochastic matrix $A \neq (1/n)J$ with $\mathrm{per}A = n!/n^n$. Choose such A with as few zero components as possible.

If at least n-1 columns of A are strictly positive, we can assume that $A = (B,x,y)$ with $B > 0$, $x > 0$ and $x \neq y$. Then from (32) it follows that we have equality in (19). Hence by Theorem 1, $y = \lambda x$ for some λ. As A is doubly stochastic, we have $\lambda = 1$ and $x = y$, contradicting our assumption.

If A has at most n-2 strictly positive columns, we can assume that $A =$ (B,x,y) is such that not all columns of B are positive, and such that y has a zero in at least one coordinate place where x is positive. Then by (34) $(B,\frac12x+\frac12y,\frac12x+\frac12y)$ is again a minimizing matrix, distinct from $(1/n)J$, but with fewer zeros than A, contradicting our choice of A.

3. VOORHOEVE'S BOUND AND BEST LOWER BOUNDS.

Erdös and Rényi [7] posed in 1968 the following conjecture, weaker than Van der Waerden's conjecture: there exists an $\varepsilon > 0$ such that if $A \in \Lambda_n^k$ with $k \geq 3$ then $\mathrm{per}A \geq (1+\varepsilon)^n$. We recall that Λ_n^k denotes the set of nonnegative integral $n\times n$-matrices with all line sums equal to k.

Erdös and Rényi's conjecture was proved independently by Voorhoeve [29] and by Bang [2] and Friedland [10]. The latter two showed that $\mathrm{per}A \ge e^{-n}$ for each doubly stochastic matrix of order n. Hence $\mathrm{per}A = k^n \mathrm{per}((1/k)A) \ge (k/e)^n$ for A in Λ_n^k. For a derivation of this result, see Section 4.

In this section we focus on Voorhoeve's result, which says that $\mathrm{per}A \ge (4/3)^n$ for each $A \in \Lambda_n^3$. This improves lower bounds found earlier considerably the best one being $\mathrm{per}A \ge 3n-2$ for $A \in \Lambda_n^3$ (Hartfiel and Crosby [11]).

The trick of Voorhoeve consists of considering the collection:

(35) $\tilde{\Lambda}_n^3$:= the collection of nonnegative integral $n\times n$-matrices with row sums $2,3,\ldots,3$ and column sums $2,3,\ldots,3$.

He showed that also for matrices A in $\tilde{\Lambda}_n^3$ one has $\mathrm{per}A \ge (4/3)^n$. This stronger result turned out to be the key to applying induction.

<u>THEOREM 4</u> (Voorhoeve's bound). *If* $A \in \Lambda_n^3$ *then* $\mathrm{per}A \ge (4/3)^n$.

<u>PROOF</u>. It is shown that $\mathrm{per}A \ge (4/3)^n$ for $A \in \tilde{\Lambda}_n^3$ by induction on n. This implies the theorem, as if $A \in \Lambda_n^3$ and B arises from A by decreasing one positive entry of A by one, then $B \in \tilde{\Lambda}_n^3$ and $\mathrm{per}A \ge \mathrm{per}B \ge (4/3)^n$.

So let $A \in \tilde{\Lambda}_n^3$. Without loss of generality the first row and the first column both have sum 2. There are the following four cases, possibly after permuting the columns of A (a,b and c denote column vectors of length n-1).

(36) $$\mathrm{per}A = \mathrm{per}\begin{pmatrix} 0 & 1 & 1 & 0\ldots0 \\ a & b & c & D \end{pmatrix} \overset{1}{=} \mathrm{per}(a,b,D)+\mathrm{per}(a,c,D) \overset{2}{=} \mathrm{per}(a,b+c,D) \overset{3}{=} \frac{1}{3}(\mathrm{per}(a,d_1,D)+\mathrm{per}(a,d_2,D)+\mathrm{per}(a,d_3,D)+\mathrm{per}(a,d_4,D)) \overset{4}{\ge} \frac{1}{3}.4\left(\frac{4}{3}\right)^{n-1} = (4/3)^n.$$

(Explanation: [1] follows by expanding the permanent by the upper row; [2] follows as the permanent is linear in the columns; [3] the components of b+c add up to 4; hence we can write $b+c = \frac{1}{3}(d_1+d_2+d_3+d_4)$ with d_1,d_2,d_3,d_4 nonnegative integral column vectors, each with column sum 3; [4] this inequality follows from the induction hypothesis, as each (a,d_i,D) belongs to $\tilde{\Lambda}_{n-1}^3$.)

(37) $$\mathrm{per}A = \mathrm{per}\begin{pmatrix} 0 & 2 & 0\ldots0 \\ a & b & D \end{pmatrix} \overset{5}{=} 2.\mathrm{per}(a,D) \overset{6}{\ge} 2\left(\frac{4}{3}\right)^{n-1} \ge \left(\frac{4}{3}\right)^n.$$

(Explanation: [5] expand the permanent by the upper row; [6] since (a,D) belongs to $\tilde{\Lambda}^3_{n-1}$, we can apply the induction hypothesis.)

(38) $$\text{per}A = \text{per}\begin{pmatrix}1 & 1 & 0\ldots0\\ 0 & 0 & D\end{pmatrix} \overset{7}{=} \text{per}(a,D)+\text{per}(b,D) \overset{8}{=} \text{per}(a+b,D) \overset{9}{=}$$
$$\tfrac{1}{2}(\text{per}(d_1,D)+\text{per}(d_2,D)+\text{per}(d_3,D)) \overset{10}{\geq} \frac{3}{2}.(\frac{4}{3})^{n-1} \geq (\frac{4}{3})^n.$$

(Explanation: [7] expand the permanent by the upper row; [8] as the permanent is linear in the columns; [9] the components of a+b add up to 3; write $a+b = \frac{1}{2}(d_1+d_2+d_3)$ with d_1,d_2,d_3 nonnegative integral vectors each with sum 2; [10] since each matrix (d_i,D) belongs to $\tilde{\Lambda}^3_{n-1}$, this inequality follows from the induction hypothesis.)

(39) $$\text{per}A = \text{per}\begin{pmatrix}2 & 0\ldots0\\ 0 & D\end{pmatrix} \overset{11}{=} 2.\text{per}D \overset{12}{\geq} 2.\text{per}D' \overset{13}{\geq} 2.(\frac{4}{3})^{n-1} \geq (\frac{4}{3})^n.$$

(Explanation: [11] expand the permanent by the upper row; [12] let D' arise from D by decreasing one positive entry of D by one; [13] since $D' \in \tilde{\Lambda}^3_{n-1}$, this inequality follows from the induction hypothesis.) □

By sharpening the method, Voorhoeve showed the better lower bound of $\frac{81}{32}(\frac{4}{3})^n$. However, the ground number 4/3 is best possible. This follows by taking $k=3$ in the following result of [26] (cf. Wilf [31]), which is proved by an averaging argument.

<u>THEOREM 5</u>. *Let* $f(k)$ *be the largest number such that* $\text{per}A \geq f(k)^n$ *for each* $A \in \Lambda^k_n$. *Then*

(40) $$f(k) \leq \frac{(k-1)^{k-1}}{k^{k-2}}.$$

<u>PROOF</u>. Let $P_{k,n}$ be the collection of all ordered partitions of $\{1,2,\ldots,nk\}$ into n classes of size k. So we have

(41) $$p_{k,n} := |P_{k,n}| = \frac{(nk)!}{k!^n}.$$

A *system of distinct representatives* (SDR) of a partition $\mathcal{A} = (A_1,\ldots,A_n)$ in $P_{k,n}$ is a subset S of $\{1,\ldots,nk\}$ such that $|S \cap A_i| = 1$ for $i=1,\ldots,n$. Clearly the number of SDR's of $\mathcal{A}$ is equal to k^n.

Now let $\mathcal{A} = (A_1,\ldots,A_n)$ and $\mathcal{B} = (B_1,\ldots,B_n)$ be in $P_{k,n}$. Let $s(\mathcal{A},\mathcal{B})$

denote the number of *common* SDR's of $\mathcal{A}$ and $\mathcal{B}$. Then $s(\mathcal{A},\mathcal{B})$ is equal to the permanent of the matrix $C = (c_{ij})_{i,j=1}^{n}$, where

(42) $$c_{ij} = |A_i \cap B_j| \qquad (i,j = 1,\dots,n).$$

Indeed, if σ is a permutation of $\{1,\dots,n\}$, then $\Pi_{i=1}^{n} a_{i\sigma(i)}$ is the number of common SDR's S containing an element in $A_i \cap B_{\sigma(i)}$, for each i. Hence

(43) $$s(\mathcal{A},\mathcal{B}) = \sum_{\sigma\in S_n} \Pi_{i=1}^{n} a_{i\sigma(i)} = \text{per}C.$$

Since $\sum_{i=1}^{n} c_{ij} = |B_j| = k = |A_i| = \sum_{j=1}^{n} c_{ij}$, we know that $C \in \Lambda_n^k$. Therefore,

(44) $$s(\mathcal{A},\mathcal{B}) \geq f(k)^n.$$

Now let $\mathcal{A} \in P_{k,n}$ be fixed. Any SDR $S = \{s_1,\dots,s_n\}$ of $\mathcal{A}$ is an SDR of $n!p_{k-1,n}$ partitions $\mathcal{B}$ in $P_{k,n}$, as we can distribute $s_1,\dots,s_n$ in n! ways among $B_1,\dots,B_n$, whereas the other elements of $B_1,\dots,B_n$ can be chosen freely. Since $\mathcal{A}$ has k^n SDR's, we find

(45) $$\sum_{\mathcal{B}\in P_{k,n}} s(\mathcal{A},\mathcal{B}) = k^n n!\, p_{k-1,n}.$$

Combining (41), (44) and (45) gives:

(46) $$f(k)^n \leq \frac{k^n n! p_{k-1,n}}{p_{k,n}} = \frac{k^n n! k!^n (nk-n)!}{(k-1)!^n (nk)!} = k^{2n}/\binom{nk}{n}.$$

By Stirling's formula, (46) implies (40). □

We conjecture that in fact the upper bound in Theorem 5 always gives the right value of f(k). This is trivially true if $k = 1$ or $k = 2$ (as $f(1) = f(2) = 1$), and is also true for $k = 3$ by Voorhoeve's theorem (Theorem 4). At the end of the following Section we shall see some more lower bounds for f(k).

Note that the proof of Theorem 5 in fact gives (cf. (46)):

(47) $$\min_{A\in\Lambda_n^k} \text{per } A \leq k^{2n}/\binom{nk}{n}.$$

4. BANG'S LOWER BOUND AND EDGE-COLOURINGS.

We now give a proof of Bang's lower bound of e^{-n} for permanents of doubly stochastic matrices of order n. His method can be interpreted, and extended, in terms of edge-colourings, or 1-factorizations, of bipartite graphs. A k-*edge-colouring* of a bipartite graph is an ordered partition of the edge set of the graph into k classes, each class being a perfect matching. It is a well-known theorem of König [13] that each k-regular bipartite graph has at least one k-edge-colouring (see Remark 1 in Section 1). Here we consider counting them.

In [24] it is shown that if $k = 2^a 3^b$, and G is a k-regular bipartite graph with 2n vertices, then

(48) G has at least $\left(\frac{k!^2}{k^k}\right)^n$ k-edge-colourings.

Moreover it is shown that for each fixed k, the ground number in (48) is best possible. It is conjectured that (48) holds for every k. This conjecture would follow from the conjecture made in Section 3 that $f(k) = (k-1)^{k-1}/k^{k-2}$ for each k, that is, that each k-regular bipartite graph has at least $\left((k-1)^{k-1}/k^{k-2}\right)^n$ perfect matchings. We could first choose a perfect matching, delete this perfect matching, next choose a perfect matching in the remainder and so on. Hence G would have at least

(49) $$\left(\frac{(k-1)^{k-1}}{k^{k-2}} \cdot \frac{(k-2)^{k-2}}{(k-1)^{k-3}} \cdot \ldots \cdot \frac{2^2}{3^1} \cdot \frac{1^1}{2^0}\right)^n = \left(\frac{k!^2}{k^k}\right)^n$$

k-edge-colourings.

In other words, let g(k) be the highest number such that each k-regular bipartite graph with 2n points has at least $g(k)^n$ k-edge-colourings. Then $g(k) \le k!^2/k^k$, and we have equality if $k = 2^a 3^b$. This is the content of the following two theorems, the first one being proved similarly to Theorem 5.

<u>THEOREM 6</u>. $g(k) \le k!^2/k^k$.

<u>PROOF</u>. Again, let $P_{k,n}$ and $p_{k,n}$ be as in the proof of Theorem 5. For $\mathcal{A}, \mathcal{B}$ in $P_{k,n}$ denote by $c(\mathcal{A},\mathcal{B})$ the number of partitions $\mathcal{C} = (C_1,\ldots,C_k)$ of $\{1,\ldots,nk\}$ into k classes of size n such that

(50) $$|A_i \cap C_j| = |B_i \cap C_j| = 1$$

for $i=1,\dots,n$ and $j=1,\dots,k$. That is, each C_j is a common SDR for $\mathcal{A}$ and $\mathcal{B}$. It is easy to see that $c(\mathcal{A},\mathcal{B})$ is equal to the number of k-edge-colourings of the k-regular bipartite graph with vertices, say, $v_1,\dots,v_n,w_1,\dots,w_n$, where v_i and w_j are connected by $|A_i \cap B_j|$ edges, for $i,j=1,\dots,n$. In particular,

(51) $$c(\mathcal{A},\mathcal{B}) \geq g(k)^n.$$

Now let $\mathcal{A} \in P_{k,n}$ be fixed. There are $k!^n$ possible partitions $\mathcal{C} = (C_1,\dots,C_n)$ of $\{1,\dots,nk\}$ with $|A_i \cap C_{j_k}| = 1$ for $i=1,\dots,n$ and $j=1,\dots,k$. For each such partition, there are $n!^k$ partitions $\mathcal{B}$ in $P_{k,n}$ such that $|B_i \cap C_j| = 1$ for $i=1,\dots,n$ and $j=1,\dots,k$. So

(52) $$\sum_{\mathcal{B} \in P_{k,n}} c(\mathcal{A},\mathcal{B}) = k!^n . n!^k.$$

Combining (41), (51) and (52) gives

(53) $$g(k)^n \leq \frac{k!^{2n} n!^k}{(nk)!} .$$

By Stirling's formula, (53) implies Theorem 6. □

A special case of the idea behind the next theorem is the following. Let $G = (V,E)$ be a 2k-regular bipartite graph, with 2n points. A *k-factor* is a collection E' of edges of G such that each point is contained in exactly k edges in E'. So E' is a k-factor in G if and only if $E \setminus E'$ is a k-factor.

Now it is easy to see that the number of k-factors of G is equal to the number $\varepsilon(G)$ of eulerian orientations of G. The latter can be seen to be at least

(54) $$\left(2^{-k}\binom{2k}{k}\right)^{2n}.$$

Indeed, we can replace the graph G by a graph G', by splitting each point v of G into k copies, and by distributing the 2k edges incident with v among the k copies of v, in such a way that G' will be 2-regular. Then G' trivially has an eulerian orientation, which induces an eulerian orientation in G. Moreover, each eulerian orientation in G arises in this way from an eulerian orientation in exactly $k!^{2n}$ graphs G' (as in each point of G we have to make pairs of an ingoing and an outgoing edge). Since there are exactly

(55) $$\left(2^{-k}\frac{(2k)!}{k!}\right)^{2n}$$

graphs G' in total, the number of eulerian orientations of G is at least (55) divided by $k!^{2n}$, which is (54).

With this it can be seen that any 2^t-regular bipartite graph G = (V,E) on 2n points has at least

(56) $$\left(\frac{(2^t)!^2}{2^{t2^t}}\right)^n$$

2^t-edge-colourings (by Theorem 6, the ground number in (56) is best possible). This can be shown by induction on t, the case t = 0 being trivial. By (54), G has at least

(57) $$\left(2^{-2^{t-1}}\binom{2^t}{2^{t-1}}\right)^{2n}$$

2^{t-1}-factors E'. By induction, the graphs (V,E') and (V,E\E') have at least

(58) $$\left(\frac{(2^{t-1})!^2}{2^{(t-1)2^{(t-1)}}}\right)^n$$

2^{t-1}-edge-colourings. So the number of 2^t-edge-colourings of G is at least (58) squared times (57), which is (56).

This idea is extended in Theorem 7.

<u>THEOREM 7</u>. *If* $g(k) = k!^2/k^k$ *for* k = s *and* k = t, *then also for* k = st.

<u>PROOF</u>. Let G = (V,E) be an st-regular bipartite graph with 2n points, with, say, $\phi(G)$ st-edge-colourings. Consider all possible graphs G' arising from G as follows. Each point of G is split into s new vertices, where each edge e of G is replaced by one new edge connecting two of the new vertices replacing the endpoints of the original edge e, in such a way that the new graph G' is t-regular. So the number of graphs G' arising in this way from G is equal to:

(59) $$\left(\frac{(st)!}{t!^s}\right)^{2n},$$

since for each point v of G we have to partition the edges incident to v into s classes of size t, which can be done in $(st)!/t!^s$ ways.

Let Π be the collection of all partitions $(E_1,\dots,E_t)$ of the edge set of G into t classes, such that each class E_j is an s-factor of G. Now any t-edge-colouring $(E_1,\dots,E_t)$ of a derived graph G' yields a partition in Π. Conversely, each partition in Π arises in this way from a t-edge-colouring of $s!^{2tn}$ graphs G' (as for each point v of G and for each j=1,...,t, we have to take care that the edges in E_j incident to v will go to distinct copies of v in G', which means that for each v and j there are s! possibilities).

Hence, by (59),

(60) $$|\Pi| \geq \left(\frac{(st)!}{t!^s}\right)^{2n} . g(t)^{sn} \Big/ s!^{2tn}$$

as each graph G' has at least $g(t)^{sn}$ t-edge-colourings.

Now each class E_j of a partition $\mathcal{E}$ in Π can be refined to an s-edge-colouring of the graph (V,E_j) in at least $g(s)^n$ ways. So $\mathcal{E}$ can be refined to an st-edge-colouring of G in at least $g(s)^{tn}$ ways. Therefore, the total number $\phi(G)$ of st-edge-colourings of G satisfies (using (60)):

(61) $$\phi(G) \geq |\Pi| . g(s)^{tn} \geq \left(\frac{(st)!}{t!^s . s!^t}\right)^{2n} . g(s)^{tn} . g(t)^{sn} = \left(\frac{(st)!^2}{(st)^{st}}\right)^n .$$

As this holds for each st-regular bipartite graph G with 2n points, it follows that $g(st) \geq (st)!^2/(st)^{st}$. □

This implies the following.

<u>COROLLARY 7a</u>. *If* k *has no other prime factors than* 2 *and* 3, *then any* k-*regular bipartite graph with* 2n *points has at least* $(k!^2/k^k)^n$ k-*edge-colourings. For fixed* k *this ground number is best possible.*

<u>PROOF</u>. By Theorems 6 and 7 it suffices to show that $g(2) \geq 1$ and $g(3) \geq 4/3$. The former inequality is trivial, while the latter follows from Voorhoeve's lower bound (Theorem 4) that the number of perfect matchings in a 3-regular bipartite graph with 2n points is at least $(4/3)^n$. □

From Theorem 7 one can also derive the lower bound of Bang [2] and Friedland [10].

COROLLARY 7b. *The permanent of a doubly stochastic matrix of order* n *is at least* e^{-n}.

PROOF. Since the dyadic doubly stochastic matrices form a dense subset of the space of all doubly stochastic matrices, it suffices to prove the lower bound for dyadic matrices only. Let $A = (a_{ij})_{i,j=1}^{n}$ be a dyadic doubly stochastic matrix. Let u be a natural number such that $2^u A$ is integral, and let for each $t \geq u$, G_t be the 2^t-regular bipartite graph with points $v_1,\ldots,v_n,w_1,\ldots,w_n$, where there are $2^t a_{ij}$ edges connecting v_i and w_j, for $i,j = 1,\ldots,n$. This means that for $t \geq u$, the graph G_t arises from the graph G_u by replacing each edge by 2^{t-u} parallel edges.

Now the number μ of perfect matchings in G_u is easily seen to be equal to:

(62) $$\mu = 2^{un}.\text{perA}.$$

Moreover, the number γ_t of 2^t-edge-colourings of G_t satisfies:

(63) $$\gamma_t \leq \mu^{2^t}.(2^{t-u})!^{2^u n},$$

since each colouring is determined by specifying 2^t perfect matchings in G_u, together with an ordering of the 2^{t-u} "copies" in G_t of each of the $2^u n$ edges of G_u. But by Corollary 7a we know:

(64) $$\gamma_t \geq \frac{(2^t)!^{2n}}{2^{t2^t n}}.$$

Combining (62), (63) and (64) gives a lower bound for perA depending on t and n, which, by Stirling's formula, tends to e^{-n} as $t\to\infty$. □

REMARK 2. Concluding we have met above the following upper and lower bounds for the functions f(k) and g(k).

(65) $$f(k) \leq \frac{(k-1)^{k-1}}{k^{k-2}},\quad g(k) \leq \frac{k!^2}{k^k},\quad f(1)=f(2)=g(1)=g(2)=1,\quad f(3)=\frac{4}{3},$$

$$f(k) \geq \frac{k}{e}, \quad g(k) \geq f(k)g(k-1) \geq f(k)f(k-1)\ldots f(1),$$

$$g(k\ell) \geq \left(\frac{(k\ell)!}{\ell!^k k!^\ell}\right)^2 . g(k)^\ell . g(\ell)^k$$

(cf. Theorem 4, 5, 6, Corollary 7b, (61)). Moreover, by methods similar to those for Theorem 7 one shows (cf. Valiant [28]):

(66) $$f(k\ell) \geq \binom{k\ell}{k}^{2/k} . \ell^{-2} . f(\ell) . g(k)^{1/k}.$$

[To prove this, we first show that each $k\ell$-regular bipartite graph G with 2n points has at least

(67) $$\left(\binom{k\ell}{k}^2 . \ell^{-2k} . f(\ell)^k\right)^n$$

k-factors. Indeed, make all possible graphs G' as in the proof of Theorem 7 (with $s = k$ and $t = \ell$). Each of these graphs has at least $f(\ell)^{kn}$ 1-factors. Each 1-factor of G' corresponds to a k-factor in G. Conversely, any fixed k-factor in G corresponds to a 1-factor in exactly

(68) $$\left(k!\frac{(k\ell-k)!}{(\ell-1)!^k}\right)^{2n}$$

graphs G' (the edges of the k-factor have to be divided among distinct points of G'). So the number of k-factors in G is at least

(69) $$\left(\frac{(k\ell)!}{\ell!^k}\right)^{2n} . f(\ell)^{kn} . \left(k!\frac{(k\ell-k)!}{(\ell-1)!^k}\right)^{-2n}$$

(using (59)), which is equal to (67).

Now we have:

(70) (the number of 1-factors in G)$^k \geq$ (the number of k-tuples of pairwise disjoint 1-factors in G) = (the number of pairs of a k-factor in G together with a k-edge-colouring of the k-factor) $\geq$ $\left(\binom{k\ell}{k}^2 . \ell^{-2k} . f(\ell)^k . g(k)\right)^n$,

which implies (66).]

Using the bounds of (65) and (66) one can derive the following bounds for f(k) and g(k) for k = 1,2,3,4,5,6,7,8,9,10:

$f(1)=1,$	$g(1)=1,$
$f(2)=1,$	$g(2)=1,$
$f(3)=\frac{4}{3},$	$g(3)=\frac{4}{3},$
$1.5 = \frac{3}{2} \leq f(4) \leq \frac{27}{16} = 1.6875,$	$g(4)=\frac{9}{4} = 2.25,$
$1.839 \approx 5/e \leq f(5) \leq \frac{256}{125} = 2.048,$	$4.139 \approx 45/4e \leq g(5) \leq 5!^2/5^5 = 4.608,$
$2.222 \approx 20/9 \leq f(6) \leq 5^5/6^4 \approx 2.411,$	$g(6) = 6!^2/6^6 \approx 11.111,$
$2.575 \approx 7/e \leq f(7) \leq 6^6/7^5 \approx 2.776,$	$28.613 \approx \frac{700}{9e} \leq g(7) \leq 7!^2/7^7 \approx 30.844,$
$2.943 \approx 8/e \leq f(8) \leq 7^7/8^6 \approx 3.142,$	$g(8) = 8!^2/8^8 \approx 96.899,$
$3.311 \approx 9/e \leq f(9) \leq 8^8/9^7 \approx 3.508,$	$g(9) = 9!^2/9^9 \approx 339.894,$
$3.679 \approx 10/e \leq f(10) \leq 9^9/10^8 \approx 3.874,$	$1250 \approx \frac{10g(9)}{e} \leq g(10) \leq 10!^2/10^{10} \approx 1316.819.$

5. BRÈGMAN'S UPPER BOUND.

It is easy to see that the *maximum* permanent of doubly stochastic matrices is 1. Similarly, the maximum permanent of matrices in Λ_n^k is k^n. However, if we go over to a further discretization, and we restrict the entries of the matrices to 0 and 1 only, less trivial upper bounds can be obtained. In 1963, Minc [20] published a conjectured upper bound (see Theorem 8 below), which was proved in 1973 by Brègman [4]. His proof is based on ideas from convex programming and on some theory of doubly stochastic matrices. Here we give the shorter proof as described in [23]. This proof uses the fact that if $t_1,\ldots,t_r$ are nonnegative real numbers, then:

$$(71) \qquad \left(\frac{t_1+\ldots+t_r}{r}\right)^{t_1+\ldots+t_r} \leq t_1^{t_1}\cdot\ \ldots\ \cdot t_r^{t_r}.$$

[This follows from the convexity of the function xlogx, by taking logarithms of both sides of (71), and by dividing these logarithms by r.]

THEOREM 8 (Brègman's upper bound). *Let* A *be a square* {0,1}*-matrix of order* n, *with* r_i *ones in row* i ($1 \leq i \leq n$). *Then*

$$(72) \qquad \operatorname{per}A \leq \prod_{i=1}^{n} r_i!^{1/r_i}.$$

PROOF. We use induction on n, the case $n=1$ being trivial. Suppose the theorem has been shown for $(n-1)\times(n-1)$-matrices. We shall prove:

(73) $$(\mathrm{perA})^{n\mathrm{perA}} \le \Big(\prod_{i=1}^{n} r_i!^{1/r_i}\Big)^{n\mathrm{perA}},$$

which implies (72).

We first give a series of (in)equalities, which we justify afterwards. The variables i, j and k range from 1 to n. Let S denote the set of all permutations σ of $\{1,\ldots,n\}$ for which $a_{i\sigma(i)} = 1$ for all $i = 1,\ldots,n$. So $|S| = \mathrm{perA}$.

(74) $$(\mathrm{perA})^{n\mathrm{perA}} \overset{1}{=} \prod_i (\mathrm{perA})^{\mathrm{perA}} \overset{2}{\le} \prod_i \Big(r_i^{\mathrm{perA}} \prod_{\substack{k \\ a_{ik}=1}} \mathrm{perA}_{ik}^{\mathrm{perA}_{ik}}\Big) =$$

$$\overset{3}{=} \prod_{\sigma\in S} \Big((\prod_i r_i).(\prod_i \mathrm{perA}_{i\sigma(i)})\Big) \le$$

$$\overset{4}{\le} \prod_{\sigma\in S} \Big((\prod_i r_i).(\prod_i (\prod_{\substack{j \\ j\ne i \\ a_{j\sigma(i)}=0}} r_j!^{1/r_j}).(\prod_{\substack{j \\ j\ne i \\ a_{j\sigma(i)}=1}} (r_j-1)!^{1/r_j-1}))\Big) =$$

$$\overset{5}{=} \prod_{\sigma\in S} \Big((\prod_i r_i).(\prod_j (\prod_{\substack{i \\ i\ne j \\ a_{j\sigma(i)}=0}} r_j!^{1/r_j}).(\prod_{\substack{i \\ i\ne j \\ a_{j\sigma(i)}=1}} (r_j-1)!^{1/r_j-1}))\Big) =$$

$$\overset{6}{=} \prod_{\sigma\in S} \Big((\prod_i r_i).(\prod_j (r_j!^{(n-r_j)/r_j}.(r_j-1)!^{(r_j-1)/(r_j-1)}))\Big) =$$

$$\overset{7}{=} \prod_{\sigma\in S} \Big(\prod_i r_i!^{n/r_i}\Big) \overset{8}{=} \Big(\prod_i r_i!^{1/r_i}\Big)^{n\mathrm{perA}}.$$

Explanation: [1] is trivial; [2] use (71) (note that r_i is the number of k such that $a_{ik}=1$ and $\mathrm{perA} = \sum_{k,a_{ik}=1}\mathrm{perA}_{ik}$); [3] the number of factors r_i equals perA on both sides, while the number of factors perA_{ik} equals the number of $\sigma \in S$ for which $\sigma(i) = k$ (this is perA_{ik} in case $a_{ik}=1$, and 0 otherwise); [4] apply the induction hypothesis to each $A_{i\sigma(i)}$ $(i = 1,\ldots,n)$; [5] change the order of multiplication; [6] the number of i such that $i \ne j$ and $a_{j\sigma(i)}=0$ is $n-r_j$, while the number of i such that $i \ne j$ and $a_{j\sigma(i)}=1$ is r_j-1 (note that $a_{j\sigma(j)}=1$, and that the equality is proved for all fixed σ and j separately); [7] and [8] are trivial. □

In particular it follows that if all row sums of A are exactly k then

(75) $$\mathrm{perA} \le \big(k!^{1/k}\big)^n.$$

It is easy to see that for fixed k the ground number here is best possible, also if we restrict ourselves to {0,1}-matrices in Λ_n^k.

6. EULERIAN ORIENTATIONS.

As a further illustration of the results and methods above, we consider eulerian orientations. For any undirected graph G = (V,E), let $\varepsilon(G)$ denote the number of eulerian orientations of G. Here an *eulerian orientation* is an orientation of the edges such that at each vertex the indegree is equal to the outdegree.

Then if G is a loopless 2k-regular graph with n vertices, the number of eulerian orientations satisfies:

$$(76) \qquad \left(2^{-k}\binom{2k}{k}\right)^n \le \varepsilon(G) \le \left(\sqrt{\binom{2k}{k}}\right)^n,$$

and moreover, for each fixed k, the ground numbers in (76) cannot be improved ([25]).

There exists a direct relation between $\varepsilon(G)$ and the permanent function. Let G = (V,E) be a graph in which each vertex v has degree deg(v) even. Let B be the incidence matrix of G, with |V| rows and |E| columns. Let the matrix A arise from B by repeating, for each vertex v, the row of B corresponding with v ½deg(v) times. Then A is a square {0,1}-matrix of order |E|. Now one easily checks that:

$$(77) \qquad \varepsilon(G) = \frac{\text{per}A}{\prod_{v\in V}(\frac{1}{2}\deg(v))!} .$$

Substituting Brègman's upper bound (Theorem 8) in (77) gives:

$$(78) \qquad \varepsilon(G) \le \prod_{v\in V}\binom{\deg(v)}{\frac{1}{2}\deg(v)}^{\frac{1}{2}},$$

and the right hand side in (76) follows. The graph with 2 points connected by 2k parallel edges shows that we cannot have a lower ground number in the upper bound in (76).

Concerning lower bounds, Falikman and Egorychev's lower bound, in the form (6), gives that if G is 2k-regular, then $A\in\Lambda_{kn}^{2k}$, and so with (77):

$$(79) \qquad \varepsilon(G) \ge \left(\frac{2}{n}\right)^{kn}\cdot\frac{(nk)!}{k!^n} .$$

Asymptotically this implies:

$$(80)\qquad \varepsilon(G) \geq \left(\frac{1}{k!}\left(\frac{2k}{e}\right)^k\right)^n.$$

The conjecture (10) would imply the better lower bound:

$$(81)\qquad \varepsilon(G) \geq \left(\frac{1}{k!}\,\frac{(2k-1)^{2k-1}}{(2k)^{2k-2}}\right)^n.$$

However, the lower bound given in (76) is even higher (and is best possible). This is not surprising, as generally the permanent function seems to approach its minimum value if the matrix tends to have a random structure, whereas the matrix A obtained from G as above, has several equal rows.

The lower bound in (76) can be shown as follows. Let $\varepsilon(2d_1,\ldots,2d_n)$ be the minimum of $\varepsilon(G)$, where G ranges over all undirected graphs (possibly with loops) with degree sequence $2d_1,\ldots,2d_n$. Then:

$$(82)\qquad \varepsilon(2d_1,\ldots,2d_n) \geq \prod_{i=1}^{n} 2^{-d_i}\binom{2d_i}{d_i}.$$

This can be seen by induction on $2d_1+\ldots+2d_n$. If this sum is 0, (82) is trivial. If, say, $d_1 \geq 1$, let G be an undirected graph with degree sequence $2d_1,\ldots,2d_n$ and with $\varepsilon(G) = \varepsilon(2d_1,\ldots,2d_n)$. Let point v have degree $2d_1$, and let $e_1,\ldots,e_{2d_1}$ be the edges incident with v. For $1 \leq i < j \leq 2d_1$, let $\varepsilon_{ij}(G)$ denote the number of eulerian orientations of G in which e_i and e_j are oriented in series (i.e., one of them has v as tail, and the other has v as head). If, say, $e_i = \{u,v\}$ and $e_j = \{v,w\}$, let G_{ij} be the graph obtained from G by replacing e_i and e_j by one new edge $\{u,w\}$. Then:

$$(83)\qquad \varepsilon_{ij}(G) = \varepsilon(G_{ij}) \geq \varepsilon(2d_1-2,2d_2,\ldots,2d_n).$$

Therefore, inductively,

$$(84)\qquad \varepsilon(G) = \frac{1}{d_1^2}\sum_{1\leq i<j\leq 2d_1} \varepsilon_{ij}(G) \geq d_1^{-2}.\binom{2d_1}{2}\varepsilon(2d_1-2,2d_2,\ldots,2d_n) \geq$$
$$\geq d_1^{-2}\binom{2d_1}{2}2^{-(d_1-1)}\binom{2d_1-2}{d_1-1}\prod_{i=2}^{n} 2^{-d_i}\binom{2d_i}{d_i} = \prod_{i=1}^{n} 2^{-d_i}\binom{2d_i}{d_i}.$$

So (82) is proved, and the lower bound in (76) follows.

By averaging techniques, similar to those in the proofs of the Theorems

5 and 6, one shows that for fixed k the ground number in the lower bound in (76) is best possible. It is also best possible if we restrict G to loopless graphs. This follows with the help of the Alexandroff-Fenchel permanent inequality (Theorem 1) - see [25]. We conjecture that it is even best possible if G is restricted to simple graphs (i.e., no loops or multiple edges). Moreover, we conjecture that for simple graphs a better upper bound can be obtained: if G is a simple undirected graph with degree sequence $2d_1,\ldots,2d_n$, then

$$(85) \qquad \text{(Conjecture)} \quad \varepsilon(G) \le \prod_{i=1}^{n} \varepsilon(K_{2d_i+1})^{1/(2d_i+1)}$$

(K_t being the complete undirected graph on t points). A problem we met in constructing a proof similar to that of Brègman's upper bound (Theorem 8) is that we could not find a suitable formula for $\varepsilon(K_t)$.

REFERENCES.

[1] A.D. Alexandroff, *K teorii smeshannykh ob'ëmov vypuklykh tel (Zur Theorie der gemischten Volumina von konvexen Körpern)*, Mat. Sb. 3 (45) (1938) 227-251 (Russian; German summary).

[2] T. Bang, *Matrixfunktioner som med et numerisk lille deficit viser v.d. Waerdens permanenthypothese*, preprint, 1978.

[3] G. Birkhoff, *Tres observaciones sobre el algebra lineal*, Rev. Univ. Nac. Tucuman Ser. A 5 (1946) 147-148.

[4] L.M. Brègman, *Some properties of nonnegative matrices and their permanents*, Soviet Math. Dokl. 14 (1973) 945-949 (English translation of: Dokl. Akad. Nauk SSSR 211 (1973) 27-30).

[5] H. Busemann, *Convex surfaces*, Interscience, New York, 1958.

[6] G.P. Egorychev, *Solution of van der Waerden's permanent conjecture*, Advances in Math. 42 (1981) 299-305 (English translation of preprint IFSO-13M of the Kirensko Institute of Physics, Krasnoiarsk, 1980).

[7] P. Erdös and A. Rényi, *On random matrices II*, Studia Sci. Math. Hungar. 3 (1968) 459-464.

[8] D.I. Falikman, *A proof of the van der Waerden conjecture on the permanent of a doubly stochastic matrix*, Mat. Zametki 29 (1981) 931-938 (Russian).

[9] W. Fenchel, *Inégalités quadratiques entre les volumes mixtes des corps convexes*, C.R. Acad. Sci. Paris 203 (1936) 647-650.

[10] S. Friedland, *A lower bound for the permanent of a doubly stochastic matrix*, Annals of Math. 110 (1979) 167-176.

[11] D.J. Hartfiel and J.W. Crosby, *On the permanent of a certain class of (0,1)-matrices*, Canad. Math. Bull. 14 (1971) 507-511.

[12] D.E. Knuth, *A permanent inequality*, Amer. Math. Monthly 88 (1981) 731-740.

[13] D. König, *Graphen und ihre Anwendung auf Determinantentheorie und Mengenlehre*, Math. Ann. 77 (1916) 453-465.

[14] D. König, *Über trennende Knotenpunkte in Graphen (nebst Anwendungen auf Determinanten und Matrizen)*, Acta Lit. Sci. Sect. Sci. Math. (Szeged) 6 (1932-1934) 155-179.

[15] J.H. van Lint, *Combinatorial Theory Seminar*, Springer Lecture Notes in Mathematics 382, Springer, Berlin, 1974.

[16] J.H. van Lint, *Notes on Egoritsjev's proof of the van der Waerden conjecture*, Linear Algebra and Appl. 39 (1981) 1-8.

[17] J.H. van Lint, *The van der Waerden conjecture: two proofs in one year*, Math. Intelligencer 4 (1982) 72-77.

[18] D. London, *Some notes on the van der Waerden conjecture*, Linear Algebra and Appl. 4 (1971) 155-160.

[19] M. Marcus and M. Newman, *On the minimum of the permanent of a doubly stochastic matrix*, Duke Math. J. 26 (1959) 61-72.

[20] H. Minc, *Upper bounds for permanents of (0,1)-matrices*, Bull. Amer. Math. Soc. 69 (1963) 789-791.

[21] H. Minc, *Permanents*, Encyclopedia of Mathematics and its Applications 6, Addison-Wesley, Reading, Mass., 1978.

[22] J. von Neumann, *A certain zero-sum two-person game equivalent to the optimum assignment problem*, in: "Contributions to the theory of games" (A.W. Tucker and H.W. Kuhn, eds.), Annals of Math. Studies

38, Princeton Univ. Press, Princeton, N.J., 1953, pp. 5-12.

[23] A. Schrijver, *A short proof of Minc's conjecture*, J. Combinatorial Theory (A) 25 (1978) 80-83.

[24] A. Schrijver, *On the number of edge-colourings of regular bipartite graphs*, Discrete Math. 38 (1982) 297-301.

[25] A. Schrijver, *Bounds on the number of eulerian orientations*, preprint, 1983.

[26] A. Schrijver and W.G. Valiant, *On lower bounds for permanents*, Indag. Math. 42 (1980) 425-427.

[27] L.G. Valiant, *The complexity of computing the permanent*, Theor. Comput. Sci. 8 (1979) 189-201.

[28] W.G. Valiant, *On an identity of T. Bang*, Report ZW 132/79, Mathematisch Centrum, Amsterdam, 1979.

[29] M. Voorhoeve, *A lower bound for the permanents of certain (0,1)-matrices*, Indag. Math. 41 (1979) 83-86.

[30] B.L. van der Waerden, *Aufgabe 45*, Jahresber. Deutsch. Math.-Verein. 25 (1926) 117.

[31] H.S. Wilf, *On the permanent of a doubly stochastic matrix*, Canad. J. Math. 18 (1966) 758-761.

ACKNOWLEDGEMENT

I like to thank Dr E. Keith Lloyd for his careful reading of the text and his many helpful comments.

REDFIELD DISCOVERED AGAIN

J. Sheehan
University of Aberdeen, Aberdeen, Scotland, AB9 2TY.

This paper is dedicated to the memory of J. Howard Redfield.

1 INTRODUCTION

Several years ago in a lecture at a British Mathematical Colloquium splinter group (Lancaster meeting, 1978) Dr. E.K. Lloyd mentioned that he had reason to believe that J. Howard Redfield had written a second (but unpublished) paper. As is well-known Redfield's first paper [17], published in 1927, was badly neglected although at least published. Attention as far as one knows was first drawn to it by Littlewood [13] in 1950. The paper was first publicized by Harary in 1960 [10]. It seemed that Redfield had anticipated most of the later discoveries in the theory of unlabelled enumeration such as Pólya's Hauptsatz, Read's Superposition Theorem, the counting of nonisomorphic graphs and the counting of self-complementary subsets of a set with respect to a group. At the end of 1981 I received an exciting letter from Lloyd stating that this second paper [18] had indeed been found. Apparently the paper had been submitted for publication in the American Journal of Mathematics on October 19th, 1940 and was rejected by the editors in a brief letter of January 7th, 1941. Harary and Robinson [12] have written a brief account of the circumstances leading up to the discovery of Redfield's second paper and happily a special edition of the Journal of Graph Theory is to be published entirely dedicated to Redfield. This edition will actually publish Redfield's second paper and will contain several commentaries on Redfield's work. Our aim in this paper is to give an interpretation of both of Redfield's papers which really should be looked at together. We choose the language of group theory since this seems most natural and helpful. Many authors have subsequently made contributions to the subject but our aim is most certainly not to give a résumé of this later work (which would be impossible) but simply to interpret Redfield's work. Harary and Palmer [11] contains a considerable bibliography and the reader is recommended to use this bibliography for reference. Especially the aim of this paper is not to decide when and by whom a certain theorem was discovered but to illuminate and pay tribute to the work of J. Howard Redfield. It is salutary to read [14], [26] to realize the difficulties faced by the mathematical historian.

We shall refer to Redfield's first paper [17] and second paper [18] as [R1] and [R2] respectively.

If a group G acts on a set D then elements $x,y \in D$ are G-equivalent if some element of G transforms x into y; thus G induces a set of equivalence classes which are counted by theorems of the Pólya-type [15]. The number of such classes can be counted using Burnside's Lemma (see Theorem 1 and [14], [26]). If we impose certain conditions on the classes that need to be counted then generalizations of Burnside's Lemma - called Pólya-type theorems - are used.

As a very simple example [2] consider the question of the number of different dice. A die is a cube, the faces of which are numbered 1,2,3,4,5,6. Let A be the set of faces and let $B = \{1,2,3,4,5,6\}$. Then obviously the number of possible dice is the number, $6!$, of one to one mappings of A into B. Now suppose we consider mappings to be equivalent if one can be transferred into the other by a rotation of the cube. This is a natural consideration since when two such equivalent dice are thrown on the table, and this is usually what one does with dice, they are indistinguishable. There are 24 rotations of the cube and so each mapping produces 24 equivalent mappings, i.e. each equivalence class contains exactly 24 members. Hence the number of classes is $6!/24$. This is in fact a simple example of the use of Burnside's Lemma.

This elementary argument does not succeed however in the following variation of the problem. Four of the vertices of the cube are coloured red and four are coloured white. How many ways can this be done up to equivalence under the rotations of the cube? The answer is 7. A simple application of Pólya's Theorem (see Section 5) solves this problem.

Redfield, [R1], [R2], then asked a seemingly more general question. Two cubes (in fact Redfield considered icosahedra but it is convenient for us to stay with cubes) can be "superposed on each other" in 24 indistinguishable ways since there are 24 rotations of the cube. Now colour, as above, the vertices of each cube (not necessarily in the same way). How many distinguishable superpositions of the cubes are there? Redfield (see Section 3) answers the question in [R2] using a very sim le generalization of his superposition theorem [R1].

Now in [R1] Redfield asked an entirely different type of question. Given a subgroup of the group of rotations of the cube how many of these distinguishable superpositions have automorphism group conjugate to this particular subgroup? In [R2], and essentially this is the main originality of the second paper, he gives, in some sense, a solution (see Section 4) to this type of problem.

2 NOTATION AND TERMINOLOGY

Let D be a non-empty finite set and let $\mathcal{S}(D)$ denote the symmetric group on D, i.e. the set of all $|D|!$ permutations of D. Suppose G is a finite group. Then a permutation representation ρ of G on D is a group homomorphism $\rho : G \to \mathcal{S}(D)$ of G into $\mathcal{S}(D)$. Since we shall only be interested in "permutation representations" we shall simply talk of "representations". For any $d \in D$, $g \in G$, $\rho(g)d$ is the image of d under $\rho(g)$. Suppose ρ and ρ^* are representations of G on D and D^* respectively. If there exists a bijection t of D onto D^* such that

$$t\rho(g)t^{-1} = \rho^*(g) \quad (g \in G)$$

then ρ and ρ^* are equivalent. Otherwise they are distinct.

Elements $d, d^* \in D$ are $\rho(G)$-equivalent if there exists $g \in G$ such that $\rho(g)d = d^*$. This is an equivalence relation on D and its equivalence classes are called orbits of ρ (or simply orbits). In the relevant literature orbits (subject to extra constraints on the action of the group) are variously called: "transitive systems [15]; "patterns" [2]; "abstract group-reduced distributions" [R1], [R2].

To avoid repetition below we fix our notation. If the reader is in any doubt as to the meaning of a particular symbol below it will refer to notation introduced in this section. Thus the set, $orb(\rho)$, of orbits of ρ will always be given by

$$orb(\rho) = \{D_i : i = 1,2,\ldots,q\}.$$

This is of course a partition of D. An arbitrary set, $\Delta(\rho)$, of orbit representatives will always be given

$$\Delta(\rho) = \{d_i : i = 1, 2, \ldots, q\} \text{ where } d_i \in D_i.$$

The representation ρ is transitive if $q = 1$.

Let $d \in D$. Then $G_\rho(d)$ denotes the stabilizer of d in ρ, i.e. if $g \in G$ then $g \in G_\rho(d)$ if and only if $\rho(g)d = d$. It is an easy exercise (see Appendix 1) to prove that if d and d^* belong to the same orbit then $G_\rho(d)$ and $G_\rho(d^*)$ are conjugate. G_ρ is <u>invariant</u>, up to conjugacy, <u>on the orbits</u>. Since we shall, in the main, be counting orbits what really interests us here is that $|G_\rho|$ is <u>constant on the orbits</u>.

The <u>character</u> χ_ρ <u>of</u> G <u>in</u> ρ is a non-negative integral function on G. For $g \in G$ we define $\chi_\rho(G)$ to be the number of elements of D fixed by $\rho(g)$. Let P be a boolean statement and write

$$\delta(P) = \begin{cases} 1 & \text{if } P \text{ is true} \\ 0 & \text{otherwise,} \end{cases}$$

so that

$$\chi_\rho(g) = \sum_{d \in D} \delta(\rho(g)d = d) .$$

The <u>unit character</u> is denoted by 1_G, i.e. $1_G(g) = 1$ for each $g \in G$. Let θ_i be a representation of G on some set $(i = 1, 2, \ldots, r)$. Then

$$(\chi_{\theta_1}, \chi_{\theta_2}, \ldots, \chi_{\theta_r}) = \frac{1}{|G|} \sum_{g \in G} \chi_{\theta_1}(g)\chi_{\theta_2}(g) \ldots \chi_{\theta_r}(g)$$

is the <u>scalar product of</u> $\chi_{\theta_1}, \chi_{\theta_2}, \ldots, \chi_{\theta_r}$.

The whole of the sequel is a consequence of, and deeply dependent on, the following theorem which simply for convenience [14], [26] we refer to as Burnside's Lemma.

<u>THEOREM 1</u> (Burnside's Lemma)

$$(\chi_\rho, 1_G) = |\mathrm{orb}(\rho)| = \frac{1}{|G|} \sum_{g \in G} \sum_{d \in D} \delta(\rho(g)d = d) .$$

<u>Proof</u>

$$\begin{aligned}
(\chi_\rho, 1_G) &= \frac{1}{|G|} \sum_{g \in G} \chi_\rho(g) \\
&= \frac{1}{|G|} \sum_{g \in G} \sum_{d \in D} \delta(\rho(g)d = d) \\
&= \frac{1}{|G|} \sum_{d \in D} \sum_{g \in G} \delta(\rho(g)d = d) \\
&= \frac{1}{|G|} \sum_{d \in D} |G_\rho(d)| \\
&= \frac{1}{|G|} \sum_{i=1}^{q} |G_\rho(d_i)||D_i|
\end{aligned}$$

$$= \frac{1}{|G|} \sum_{i=1}^{q} |G| \quad \text{(see Appendix 2)}$$

$$= |\mathrm{orb}(\rho)| \ .$$

Commentary

In [R1] Redfield proves a corollary to Theorem 1. He also poses a question in this paper to which he gives only a partial answer. This question is then answered (in some sense) completely in [R2] . The most natural way to present, and answer Redfield's question is in the language of group theory. However we must still develop our notation a little further. We would like this paper to be reasonably self-contained but to give all proofs would take too much space. The sort of proofs omitted are no harder, and no less standard, than those given in the appendix. The reader is referred to [1] , [9] to fill in any gaps. It is important to remember that the sort of proofs given in the Appendix and elsewhere were often developed independently by Redfield.

Let H be a subgroup of G and $G{:}H$ its set $\{Hx : x \in G\}$ of right cosets with respect to G . Let ρ_H denote the representation of G on $G{:}H$ defined by right multiplication, i.e.

$$\rho_H(g)(Hx) = Hxg^{-1} \quad (x, g \in G) \ .$$

Then ρ_H is clearly transitive. Moreover it is "canonically transitive". Let us explain. Choose $G_1, G_2, \ldots, G_N$ to be a complete set of non-conjugate subgroups of G ordered so that $|G_1| \geqslant |G_2| \geqslant \ldots \geqslant |G_N|$. Instead of ρ_{G_i} and $\chi_{\rho_{G_i}}$ write simply ρ_i and χ_i $(i = 1, 2, \ldots, N)$. Then [1, p.288]

$$\{\rho_i : i = 1, 2, \ldots, N\}$$

is a complete set of distinct transitive representations. Thus (see Appendix 3) suppose θ is any transitive representation of G on a set S . Let $s \in S$. Then θ is equivalent to ρ_j , for some j, $1 \leqslant j \leqslant N$, i.e. $G_\theta(s)$ is conjugate to G_j . Now, returning to our earlier notation, the restriction of the action of ρ to D_i induces a transitive representation ψ_i of G on D_i , i.e. $\psi_i : G \to \mathcal{S}(D_i)$ is defined by

$$\psi_i(g)d = \rho(g)d \quad (d \in D_i, \ g \in G) \ .$$

Therefore ψ_i is equivalent to ρ_j and $G_{\psi_i}(d_i)$ is conjugate to G_j for some j, $1 \leqslant j \leqslant N$, $i = 1, 2, \ldots, q$. Therefore, unless we are applying our results to particular problems, we may identify D_i with $G{:}G_j$.

3 THE REDFIELD-READ SUPERPOSITION THEOREM

Suppose H is a subgroup of G. Let $\mathrm{con}(g)$ denote the conjugacy class of G to which g belongs. Then $J(H,g)$ is defined by

$$J(H,g) = H \cap \mathrm{con}(g) .$$

We include the proof, though standard [9, p.284], of the following proposition because it is proved in both [16] and [R1], and illustrates very well the ingenuity of both authors' arguments.

PROPOSITION 2 (Technical) For each $g \in G$

$$\chi_i(g) = \{|G|.|J(G_i,g)|\}/\{|G_i|.|J(G,g)|\} .$$

Proof

Suppose $G:G_i = \{G_i x_j : j = 1, 2,\ldots,m\}$.

$$\text{Then} \quad \chi_i(g) = \sum_{j=1}^{m} \delta(x_j g x_j^{-1} \in G_i) \quad (g \in G)$$

$$= \frac{1}{|G_i|} \sum_{y \in G} \delta(ygy^{-1} \in G_i) .$$

This summation counts only those $ygy^{-1} \in J(G_i,g)$ and each such element is counted exactly z times where z is the order of the centralizer of g in G. But $z = |G|/|J(G,g)|$. This completes the argument.

Let ρ_i^* be any representation of G on a set A_i $(i = 1, 2,\ldots,q)$. Let A be the cartesian product $A_1 \times A_2 \times \ldots \times A_q$ and let ρ^* be the representation of G on A defined by

$$\rho^*(g)(a_1,a_2,\ldots,a_q) = (\rho_1^*(g)a_1,\rho_2^*(g)a_2,\ldots,\rho_q^*(g)a_q)$$

$(g \in G,\ a_i \in A_i,\ i = 1, 2,\ldots,q)$. The representation ρ^* is sometimes called [8] the Kronecker product of the representations $\rho_1^*, \rho_2^*,\ldots,\rho_q^*$. Now there is no loss of generality in assuming that ρ_i^* is transitive (otherwise simply increase the number of components in the Kronecker product as in the example below) and also in assuming (up to equivalence) that

$$\rho_i^* = \rho_{j_i} , \quad A_i = D_{j_i} ,$$

where $j_i \in \{1, 2,\ldots,N\}$ $(i = 1, 2,\ldots,q)$. As in Section 2 we may identify D_{j_i} with $G:G_{j_i}$.

Example

Let $D = \{a, b, c, d, e, f\}$ and G be the permutation group acting on D defined by

$$G = \{1,(ab),(cd),(ab)(cd),(ab)(ef),(cd)(ef),(ab)(cd)(ef),(ef)\}.$$

Let $A_1 = \{a,b\}$, $A_2 = \{c, d, e, f\}$ and $A = A_1 \times A_2$. Let ρ_i^* be the representation of G on A_i induced by the action of G $(i = 1, 2)$, e.g. if $g = (ab)$ then $\rho_1^*(g)a = b$, $\rho_1^*(g)b = a$, $\rho_2^*(g)c = c$, $\rho_2^*(g)d = d$, $\rho_2^*(g)e = e$, $\rho_2^*(g)f = f$. By definition ρ^* is the representation of G on A defined by

$$\rho^*(g)(a_1,a_2) = (\rho_1^*(g)a_1, \rho_2^*(g)a_2) \quad (a_i \in A_i).$$

So in particular if $g = (a,b)$ and $a_1 = a$, $a_2 = f$

$$\rho^*(g)(a,f) = (b,f).$$

Of course ρ_1^* is transitive and ρ_2^* is intransitive.

Now let $A_1 = \{a,b\}$, $A_2 = \{c,d\}$, $A_3 = \{e,f\}$ and $A = A_1 \times A_2 \times A_3$. Let ρ_i^* be the representation of G on A_i induced by the action of G $(i = 1, 2, 3)$. Then the representation ρ^* is defined by

$$\rho^*(g)(a_1, a_2, a_3) = (\rho_1^*(g)a_1, \rho_2^*(g)a_2, \rho_3^*(g)a_3).$$

This time however each of the constituents ρ_i^* of ρ^* is transitive.

COROLLARY 3 (Redfield-Read Superposition Theorem [R1], [16])

$$|\mathrm{orb}(\rho^*)| = (\chi_{j_1}, \chi_{j_2}, \ldots, \chi_{j_q})$$

where this scalar product may be evaluated by using Proposition 2.

Proof

Since clearly, for $g \in G$,

$$\chi_{\rho^*}(g) = \chi_{j_1}(g)\chi_{j_2}(g)\ldots\chi_{j_q}(g),$$

the corollary follows directly from Theorem 1.

Commentary

In [R2] G is called the frame group. Actually Corollary 3 was proved initially in [R1] only for the case when G is a symmetric group. In [R2] no such restriction is made. At first sight we may seem some distance from Redfield's work but the difference is simply notational. One observation which needs to be emphasized immediately is that the distinction Redfield draws between "permutation groups" [R1] and "abstract groups"

[R2] is irrelevant. In both papers we are essentially only concerned with the representations $\{\rho_i : i = 1, 2, \ldots, N\}$. Accordingly we talk only of a group-reduced distribution (G.R.D. [R1]) rather than distinguishing such distributions from abstract group reduced distributions (A.G.R.D.'s [R2]). There is however a slight notational difference between [R1] and [R2] which will become apparent as we now interpret Redfield's notation in our language.

Write $H_i = G_{j_i}$. Then H_i is a range group ($i = 1, 2, \ldots, q$). Consider a typical orbit $P(\rho^*)$ of ρ^*, e.g.

$$P(\rho^*) = \{(H_1x_1g, H_2x_2g, \ldots, H_qx_qg) : g \in G\}$$

where $x_i \in G$, $i = 1, 2, \ldots, q$.

Suppose $G = \{g_i : i = 1, 2, \ldots, m\}$. Let h_i be an element of H_i for $i = 1, 2, \ldots, q$ and let B be the $q \times m$ matrix $[b_{ij}]$ where $b_{ij} = h_ix_ig_j$. Let $B(g)$ denote the $q \times m$ matrix obtained from B by right multiplication of its columns by g, i.e. the (i,j)-th element of $B(g)$ is $h_ix_ig_jg$. Then $\{B(g) : g \in G\}$ is called a range-correspondence. The set of all such range-correspondences, associated with $P(\rho^*)$, as the choice of $h_i \in H_i$ varies over H_i, is called a group-reduced distribution. This description provides a natural bijection between the set of orbits of ρ^* and the set of group-reduced distributions.

The notation in [R1] is slightly different. Here $G = \mathfrak{S}(D)$ and $D = \{1, 2, \ldots, n\}$. Now let $B = [b_{ij}]$ be the $q \times n$ matrix with terms

$$b_{ij} = (h_ix_i)(j) .$$

Suppose $g \in G$. Let $B(g)$ be the matrix obtained from B by a permutation of its columns induced by g. Precisely, the (i,j)-th element of $B(g)$ is $(h_ix_i)g^{-1}(j)$ ($i = 1, 2, \ldots, q$; $j = 1, 2, \ldots, n$). The connection is clear. Let B be a matrix (in the sense of [R1]). We may identify B with the vector $(h_1x_1, h_2x_2, \ldots, h_qx_q)$. Let $\langle B \rangle$ denote the set of $n!$ matrices obtained from B by column permutations. For example if g_i is such a permutation then we may identify the matrix so obtained by $(h_1x_1g_i, h_2x_2g_i, \ldots, h_qx_qg_i)$. The correspondence between $\langle B \rangle$ and the matrix B (in the sense of [R2]) is now established - these vectors, as g_i ($i = 1, 2, \ldots, n!$) runs through the elements of $G = \mathfrak{S}(D)$, are the columns of B (in the sense of [R2]). Now the definitions of range-correspondences and group-reduced distributions are as above.

4 REDFIELD'S QUESTION AND ITS ANSWER

In [R2] Redfield poses the following problem. "Two equal regular icosahedra are congruently superposable on each other in 60 ways, since the group of rotations of such a figure is of order 60. If the icosahedra are not marked in any way, these 60 superpositions are indistinguishable ... . Now suppose that one icosahedron has two antipodal faces marked A and that the other icosahedron has two antipodal faces marked B ... . Find the number of distinguishable superpositions." The answer is 3. This can be deduced from Corollary 3 as follows. The group of rotations of an icosahedron is isomorphic to the alternating group $\mathcal{A}_5$ on 5 symbols. So let $G = \mathcal{A}_5$. Notice that G contains 1 element of order 1, 15 conjugate elements of order 2 and 20 conjugate elements of order 3. Either marked polyhedron admits 3 rotations about the line through the centres of its marked faces and 3 rotations in which the faces are interchanged. So the group of rotations of the marked polyhedron is isomorphic to the symmetric group $\mathcal{S}_3$ on 3 symbols. Take for range groups $H_1 = H_2 = \mathcal{S}_3$. Then, from Corollary 3,

$$|\mathrm{orb}(\rho^*)| = \frac{60^2}{60.6.6}\left(\frac{1.1.1}{1^2} + \frac{15.3.3}{15^2} + \frac{20.2.2}{20^2}\right)$$

$$= 3 .$$

Now, apart from the unnecessary restriction of G to $\mathcal{S}_n$, a symmetric group on n symbols, this problem can be solved using the techniques of Redfield's first paper. However he goes on in [R2] to ask another question (and this is the question to which we now address ourselves):

"Every one of these 3 superpositions, combining as it does the markings of both H_1 and H_2 in a fixed relation, is itself a marked polyhedron having its own group of rotations." What are these rotation groups?

In order to pose this question precisely and to give Redfield's answer we must develop our notation a little further. We return to the notation of Section 2.

$$\mathrm{aut}(G_j) = \sum_{i=1}^{q} \delta(G_\rho(d_i) \sim G_j)$$

where $\sim$ denotes conjugacy in G. Write

$$a_j = \sum_{i=1}^{q} \delta(\psi_i \text{ is equivalent to } \rho_j)$$

where, as in Section 2, ψ_i is the representation of G on D_i induced by ρ.

It follows, again using the arguments of Section 2, that

$$a_j = \text{aut}(G_j) \quad (j = 1, 2, \ldots, N) .$$

Commentary

This notation is deliberately chosen, for $\text{aut}(G_j)$ is the number of "orbits of ρ with automorphism group (= stabilizer group) conjugate to G_j in G". So in Redfield's example it is the number of marked polyhedra having as their "own group of rotations" a group which is conjugate in G to G_j.

In general Redfield asked

* How can the numbers a_j be determined? *

In [R1] a partial solution is given when G is a symmetric group and the range groups are restricted to being either symmetric or cyclic. Redfield gives his complete solution in [R2]. We present it in our language (or rather in the language of [1]).

It is not difficult to prove (see [1]) that ρ is determined uniquely, up to equivalence, by the numbers a_i. Hence the formal notation

$$\rho = \sum_{i=1}^{N} a_i \rho_i$$

is meaningful, as well as being helpful.

Let H be any subgroup of G. The <u>mark</u> $M_\rho(H)$ <u>of</u> H <u>in</u> ρ is defined by

$$M_\rho(H) = \sum_{d \in D} \delta(\rho(H)d = d)$$

where $\rho(H)d = d$ means $\rho(h)d = d \;\; (\forall h \in H)$. Thus $M_\rho(H) = \chi_\rho(h)$ if H is cyclic and generated by h. If ρ is the transitive representation ρ_H (see Section 2) then for brevity we write M_H rather than M_{ρ_H}. Notice that

$$M_K(H) = \frac{1}{|K|} \sum_{g \in G} \delta(gHg^{-1} \subset K) , \qquad (1)$$

and that M_K is constant on conjugate subgroups of G. Let $M = [m_{ij}]$ be the $N \times N$ matrix defined by

$$m_{ij} = M_{G_j}(G_i) , \quad i,j = 1, 2, \ldots, N .$$

From (1), $m_{ij} > 0$ if and only if G_i is contained either in G_j or in some conjugate of G_j. Because of the ordering we have imposed on the subgroups G_i this means $m_{ij} = 0 \; (j > i)$. Furthermore $m_{ii} \geq 1$

$(i,j = 1, 2,\ldots,N)$. Hence, and crucially, M is non-singular. Finally, and again it is easy to prove,

$$M_\rho(G_j) = \sum_{i=1}^{N} a_i m_{ji} \quad (j = 1, 2,\ldots,N) . \tag{2}$$

We can now give the answer to Redfield's general question.

Solution

(a) Determine, by inspection, M and $M_\rho(G_j)$, $j = 1,2,\ldots,N$.

(b) Determine the numbers a_i from (2) , by inverting M .

For example consider the solution of Redfield's particular problem. We have $G = \mathcal{A}_5$ and a complete set $\{G_j : j = 1, 2,\ldots,9\}$ of non-conjugate subgroups of G can be described as follows:

$G_1 = G$;

G_2 = tetrahedral group of order 12;

G_3 = dihedral group of order 10;

G_4 = non-cyclic group of order 6;

G_5 = group of order 5;

G_6 = non-cyclic group of order 4;

G_7 = group of order 3;

G_8 = group of order 2;

G_9 = identity group.

By inspection M is the matrix below. The range groups are both conjugate to G_4 . By definition of ρ , $M_\rho(G_j) = M_{G_4}(G_j).M_{G_4}(G_j)$ $(j = 1, 2,\ldots,9)$. So the 1×9 matrix $[M_\rho(G_j)]$ is the inner product of the 4th column with itself, i.e.

$$[0 \quad 0 \quad 0 \quad 1 \quad 0 \quad 0 \quad 1 \quad 4 \quad 100]$$

which, by inspection, is equal to the vector sum of the 4th, 8th and 9th columns. Thus, from (2) ,

$$a_4 = a_8 = a_9 = 1 \ ; \quad a_i = 0 \ \text{ otherwise.}$$

So the answer to Redfield's question is that the three groups of rotations are G_4, G_8 and G_9 . Incidentally observe how the scalar product (see Corollary 3) can be evaluated directly without using the technical Proposition 3. In fact $M_{\rho_4}(G_9)$, $M_{\rho_4}(G_8)$ and $M_{\rho_4}(G_7)$ are the values of

$\chi_4(g_1)$, $\chi_4(g_2)$ and $\chi_4(g_3)$ where g_1, g_2 and g_3 are elements of order 1, 2 and 3 respectively. Hence, using simply χ_4, we have

$$|\text{orb}(\rho^*)| = \frac{1}{60}(1.10.10 + 15.2.2 + 20.1.1) = 3 .$$

The "matrix of characters" is a submatrix of M . However the matrix of characters is not invertible and this is why the extra information provided by the "marks" is needed.

5 REDFIELD AND PÓLYA

It is often remarked (see [10]) that Redfield's first paper implied a knowledge of Pólya's Theorem (Theorem 6 below). There can be little doubt that he "believed Pólya's Theorem to be true" and quite probably could have given a general statement of the theorem. He could have laid claim to being the first person to substitute into a cycle index. This having been said there seems to be no indication in either paper that he had the means to prove the theorem. There seems to have been one crucial step in the argument which might have escaped him and we would like to suggest below which step it was.

The crucial distinction between the Redfield-Read Theorem and Pólya-type theorems is that the former counts the number of certain equivalence classes whereas the latter produces a generating function, the

	ρ_1	ρ_2	ρ_3	ρ_4	ρ_5	ρ_6	ρ_7	ρ_8	ρ_9
G_1	1	0	0	0	0	0	0	0	0
G_2	1	1	0	0	0	0	0	0	0
G_3	1	0	1	0	0	0	0	0	0
G_4	1	0	0	1	0	0	0	0	0
G_5	1	0	1	0	2	0	0	0	0
G_6	1	1	0	0	0	3	0	0	0
G_7	1	2	0	1	0	0	2	0	0
G_8	1	1	2	2	0	3	0	2	0
G_9	1	5	6	10	12	15	20	30	60

coefficients of which are those numbers produced by the Redfield-Read Theorem. After this, and more interestingly perhaps, we show, using the language of group theory, how Redfield's work is related to Pólya's Theorem. Again we remind the reader that the notation, unless we declare to the contrary, is the notation of Sections 2 and 3.

LEMMA 4

Let w be any function defined on D which is constant on each orbit D_i of ρ. Then

$$\sum_{i=1}^{q} w(d_i) = \frac{1}{|G|} \sum_{g\in G} \sum_{d\in D} w(d)\delta(\rho(g)d = d) .$$

Proof

$$\begin{aligned}\sum_{g\in G} \sum_{d\in D} w(d)\delta(\rho(g)d = d) &= \sum_{d\in D} w(d) \sum_{g\in G} \delta(\rho(g)d = d)\\ &= \sum_{d\in D} w(d)|G_\rho(d)|\\ &= \sum_{i=1}^{q} w(d_i)|D_i|.|G_\rho(d_i)|\\ &= \sum_{i=1}^{q} w(d_i)|G|\\ &= |G| \sum_{i=1}^{q} w(d_i) .\end{aligned}$$

Let R be a finite set and $\mathfrak{F} = R^D$, i.e. $\mathfrak{F}$ is the set of functions from D into R. Let ρ^R be the representation of G on $\mathfrak{F}$ defined by

$$(\rho^R(g)f)(d) = f(\rho(g)d) \quad (\forall d \in D)$$

where $g \in G$, $f \in \mathfrak{F}$.

In this notation [3] elements of $\mathrm{orb}(\rho^R)$ are called patterns. Let $\mathcal{A}$ be a commutative algebra over $\mathbb{Q}$ - usually an algebra of rational polynomials. An element of $\mathcal{A}^R$ is called a weight function. Let w be a fixed weight function. Suppose $f \in \mathfrak{F}$. Then the weight $W(f)$ of f is defined by

$$W(f) = \prod_{d\in D} w(f(d)) .$$

Because of the commutativity of $\mathcal{A}$, W is constant on each pattern. Then the weight $W(F)$ of a pattern F is defined to be $W(f)$ for any $f \in F$.

LEMMA 5 (Technical)

Let $\mathcal{J} = \{f \in \mathcal{F} : f$ is constant on each orbit D_i of $\rho\}$.

Then

$$\sum_{f\in\mathcal{J}} W(f) = \prod_{i=1}^{q} \sum_{r\in R} (w(r))^{|D_i|} .$$

Proof

This is a simple counting argument independent of the action of G (except of course in so far as it determines $\mathrm{orb}(\rho)$). Let $f \in \mathcal{J}$. Since f is constant on each D_i, if $f(d_i) = r_i$ $(i = 1,2,\ldots,q)$, then

$$W(f) = \prod_{i=1}^{q} (w(r_i))^{|D_i|} .$$

Conversely any such product, where the elements r_i are chosen arbitrarily from R (remember the r_i need not be distinct), will be equal to $W(f)$ for some $f \in \mathcal{J}$. Hence

$$\sum_{f\in\mathcal{J}} W(f) = \sum \prod_{i=1}^{q} (w(r_i))^{|D_i|}$$

where the summation is over all possible choices of $(r_1, r_2,\ldots,r_q) \in R^q$. The result follows after rewriting this sum of products as a product of sums.

Let $g \in G$. Elements $d, d^* \in D$ are $\rho(g)$-equivalent if $\rho(g^m)d = d^*$, for some integer m. This is an equivalence relation on D, the equivalence classes are called $\rho(g)$-orbits and $\mathrm{orb}(\rho(g))$ denotes the set of $\rho(g)$-orbits. Write

$$\mathrm{orb}(\rho(g)) = \{D_i(g) : i = 1, 2,\ldots,q(g)\} .$$

Now $\rho(g)$ is of type

$$(j(g)) = (j_1(g), j_2(g),\ldots,j_{|D|}(g))$$

if there exist exactly $j_i(g)$ $\rho(g)$-orbits of size i $(i = 1, 2,\ldots,|D|)$. The cycle index $z(\rho(g) : x_1, x_2,\ldots,x_{|D|})$ of $\rho(g)$, where $x_1,x_2,\ldots,x_{|D|}$ are indeterminates, is defined by

$$z(\rho(g) : x_1,x_2,\ldots,x_{|D|}) = \prod_{i=1}^{|D|} x_i^{j_i(g)} .$$

The cycle index of G on ρ is defined by

$$Z_\rho(G : x_1,x_2,\ldots) = \frac{1}{|G|} \sum_{g\in G} z(\rho(g): x_1,x_2,\ldots) .$$

As a convenience it is often useful not to specify the number of indeterminates.

THEOREM 6 (Pólya [15])

$$\sum W(F) = Z_\rho(G : y_1, y_2, \ldots) ,$$

where the summation is over all patterns $F \in orb(\rho^R)$, and

$$y_i = \sum_{r \in R} (w(r))^i .$$

Proof

From Lemma 4 ,

$$\sum W(F) = \frac{1}{|G|} \sum_{g \in G} \sum_{f \in \mathcal{F}} W(f).\delta(\rho^R(g)f = f)$$

$$= \frac{1}{|G|} \sum_{g \in G} \sum W(f) \qquad (1)$$

where the inner summation is over the set $\mathcal{J}(g)$ consisting of all $f \in \mathcal{F}$ such that $\rho^R(g)f = f$.

Now suppose $f \in \mathcal{J}(g)$ and $d \in D$; then

$$f(d) = f(\rho(g^{-1})d) = f(\rho(g^{-2})d) = \ldots .$$

Since $\{d, \rho(g^{-1})d, \rho(g^{-2})d, \ldots\}$ is a $\rho(g)$-orbit it follows that f is constant on each orbit of $\rho(g)$. The converse is clearly true also. Hence, from Lemma 5, with $\mathcal{J}(g)$ and $\rho(g)$ playing the roles of $\mathcal{J}$ and ρ respectively,

$$\sum_{f \in \mathcal{J}(g)} W(f) = \prod_{i=1}^{q(g)} \sum_{r \in R} (w(r))^{|D_i(g)|}$$

$$= \left[\sum_{r \in R} w(r)\right]^{j_1(g)} \left[\sum (w(r))^2\right]^{j_2(g)} \ldots$$

$$= z(\rho(g) : y_1, y_2, \ldots) , \qquad (2)$$

where $y_i = \sum_{r \in R} (w(r))^i$.

The result follows from (1) and (2) .

Commentary

In [R1], Redfield provides a solution to the following problem. It is mainly his solution to this problem which suggests he was aware of the theorem which later came to be known as Pólya's Theorem.

We quote from [R1] "required the number of distinct configurations which can be obtained by placing a red color at each of four vertices of a cube, and a white color at each of the four remaining vertices, configurations differing only in orientation not being regarded as distinct".

The rotation group G_1 of the cube is isomorphic to $\mathcal{S}_4$. Let D be the set of vertices of the cube and ρ_1 be the representation of G_1 on D. Then

$$Z_{\rho_1}(G_1 : x_1,x_2,\ldots) = \frac{1}{24}(x_1^{\,8} + 9x_2^{\,4} + 8x_3^{\,2}x_1^{\,2} + 6x_4^{\,2}) \,. \quad (3)$$

Now Redfield attacks the problem from two directions. Firstly he uses Corollary 3. Let G_2 be the group which "rotates the 8 coloured vertices". So G_2 is isomorphic to the direct product of $\mathcal{S}_4$ with itself. Its representation ρ_2 on the set of vertices has cycle index

$$Z_{\rho_2}(G_2 : x_1,x_2,\ldots) = \left[\frac{1}{24}(x_1^{\,4} + 6x_2x_1^{\,2} + 3x_2^{\,2} + 8x_3x_1 + 6x_4)\right]^2. \quad (4)$$

The frame group G is $\mathcal{S}(D)$ when D is the set of vertices. Thus, from Corollary 3,

$$|\mathrm{orb}(\rho^*)| = \frac{8!}{24.576}\left(1.1 + \frac{9.9.2^4.(4!)}{8!} + \frac{64.8.3^2.1^2.(2!)^2}{8!} + \frac{36.6.4^2.2!}{8!}\right) \quad (5)$$
$$= 7 \,.$$

Secondly he uses a Pólya-type attack on the problem. Accordingly in (3) he puts $x_i = \alpha^i + \beta^i$ and correctly claims that the coefficient of $\alpha^t\beta^{8-t}$ is the number of distinct configurations with t red vertices and 8-t white vertices. Thus (3) becomes

$$\alpha^8 + \alpha^7\beta + 3\alpha^6\beta^2 + 3\alpha^5\beta^3 + 7\alpha^4\beta^4 + 3\alpha^3\beta^5 + 3\alpha^2\beta^6 + \alpha\beta^7 + \beta^8$$

and indeed the coefficient of $\alpha^4\beta^4$ is 7.

Redfield however offers no explanation as to why the two methods give the same result.

We tentatively suggest that he "knew why" but could not prove it and further suggest that the reason he could not prove it was that Lemma 5 escaped him. All the other components of Pólya's argument are contained in [R1] except for this technical lemma. Although seemingly very simple it does seem to be the nub of Pólya's argument.

Finally, in the rest of this paper, we discuss why the two results are the same.

We begin by presenting a more general version [15] of Pólya's Theorem. We retain the notation of this section but also we lean heavily on the notation of Section 2. Let $\chi^{(i)}$ denote the character of G in ψ_i. Let $(R_1, R_2, \ldots, R_q)$ be a partition of R. Write $\mathfrak{F}_i = R_i^{D_i}$ and

$$\mathfrak{F} = \mathfrak{F}_1 \times \mathfrak{F}_2 \times \ldots \times \mathfrak{F}_q .$$

Let ρ^R be the product of the representations $\psi_i^{R_i}$ of G on $\mathfrak{F}_i$ $(i = 1,2,\ldots,q)$, i.e. if $f_i \in \mathfrak{F}_i$, $\alpha_i \in D_i$, $g \in G$ then

$$(\rho^R(g)(f_1, f_2, \ldots, f_q))(\alpha_1, \alpha_2, \ldots, \alpha_q)$$

$$= (f_1(\psi_1(g^{-1})\alpha_1), \ldots, f_q(\psi_q(g^{-1})\alpha_q)) .$$

Again let $\mathcal{A}$ be a commutative algebra over $\mathbb{Q}$. Choose $w_i \in \mathcal{A}^{R_i}$. Let $f = (f_1, f_2, \ldots, f_q) \in \mathfrak{F}$. Then we define $W(f)$ by

$$W(f) = \prod_{i=1}^{q} \prod_{d \in D_i} w_i(f_i(d)) .$$

Again, because of the commutativity of $\mathcal{A}$, W is constant on patterns and for any pattern F we define $W(F) = W(f)$ for any $f \in F$.

<u>THEOREM 7</u> (Pólya [15])

$$\sum W(F) = \frac{1}{|G|} \sum_{g \in G} \prod_{i=1}^{q} z(\psi_i(g) : y_{i1}, y_{i2}, \ldots) ,$$

where the summation is over all patterns $F \in \text{orb}(\rho^R)$, and

$$y_{ji} = \sum_{r \in R_j} (w_j(r))^i .$$

<u>The Final Link</u>

The Redfield-Read Theorem (Corollary 3) can easily be deduced from Theorem 7. Let $R_i = \{0,1\}$ and $w_i(0) = 1$, $w_i(1) = t_i$. Suppose $\psi_i(g)$ is of type $(j_1^{(i)}, j_2^{(i)}, \ldots)$, $i = 1,2,\ldots,q$. We should here indicate that $j_k^{(i)}$ depends on g but enough is enough. We have, from Theorem 7,

$$|\text{orb}(\rho^*)| = \mathcal{C}(t_1 t_2 \ldots t_q)\{\frac{1}{|G|} \sum_{g \in G} \prod_{i=1}^{q} z(\psi_i(g) : y_{i1}, y_{i2}, \ldots)\}$$

where $y_{ji} = 1 + t_j^{\,i} (j = 1,2,\ldots,q)$, ρ^* is the Kronecker product of the induced representations of G on D_i described in Section 3 and $\mathcal{C}(t_1t_2\ldots t_q)\ \{\ldots\}$ denotes the coefficient of $t_1t_2\ldots t_q$ in the expression within the braces. Therefore

$$|\mathrm{orb}(\rho^*)|$$

$$= \mathcal{C}(t_1t_2\ldots t_q)\{\frac{1}{|G|}\sum_{g\in G}\prod_{i=1}^{q}((1+t_i)^{j_1^{(i)}}\ (1+t_i^{\,2})^{j_2^{(i)}}\ \ldots)\}$$

$$= \frac{1}{|G|}\sum_{g\in G}(\prod_{i=1}^{q} j_1^{(i)}(g))$$

$$= (\chi^{(1)}, \chi^{(2)},\ldots,\chi^{(q)}) .$$

Using the rearrangement in Section 3 this becomes

$$|\mathrm{orb}(\rho^*)| = (\chi_{j_1},\chi_{j_2},\ldots,\chi_{j_q}) .$$

So Corollary 3 is deduced from Theorem 7.

In our earlier group-theoretic language we can interpret this result in the following way. Let F be a pattern and $f = (f_1,f_2,\ldots f_q) \in F$. Suppose $W(F) = W(f) = t_1t_2\ldots t_q$. Then, since 1 and t_i are the weights of 0 and 1 respectively (under the mapping w_i) , $w_i(f_i) = t_i$. But $w_i(f_i) = \prod_{d\in D_i} w_i(f_i(d))$. Hence there exists $d_i^{\,*} \in D_i$ such that $f_i(d_i^{\,*}) = 1$ and $f_i(d) = 0\ (\forall\, d \in D_i \setminus \{d_i^{\,*}\})$. Now write $H_i = G_{j_i}$ $(i = 1,2,\ldots,q)$. Then (see Section 2) we may suppose there exists $x_i \in G$ such that $d_i^{\,*} = H_ix_i$. In this way there exists a natural mapping between the set of patterns $\overline{F}$ counted by Pólya's Theorem and the orbits, $\{(H_1x_1g, H_2x_2g,\ldots,H_qx_qg) : g \in G\}$, of ρ^* .

<u>Commentary</u>

As far as one knows attention to Theorem 7 and its link with the Redfield-Read Superposition Theorem was first drawn in [19], [20], [21], [22] . Theorem 7 is proved in [20]. However no credit in this case goes to the author since on rereading [15] he noticed that Pólya himself states the same theorem together with the aside "let us quickly point out a generalization [i.e. Theorem 7] ... which may be used in related questions". In other words Pólya believed it to be trivially self-evident.

There are many generalizations of Pólya's Theorem and many of these generalizations can be deduced from the theorems of which they are generalizations. For almost the final word on this subject the reader is recommended to read [4], [5], [6] in which the "monster generalization" is stated. However it is in the particular applications of these theorems, shown by such as Redfield, where the interest and ingenuity often lies rather than in the hierarchy of theorems to which they belong.

The results of Section 4 are all of course contained in [1]. In this context attention was first drawn to them in [7], [8], [22]. Later authors such as White in a delightful paper [24], Williamson [25] and Stockmeyer [23] also used those ideas. Again we make no attempt to give a comprehensive bibliography, which would be difficult.

<u>Appendix 1</u> (see page 137)

<u>Proof</u> If $d, d^* \in D_i$ then $\rho(g)d = d^*$ for some $g \in G$. Hence $gG_\rho(d)g^{-1} = G_\rho(d^*)$ since if $x \in gG_\rho(d)g^{-1}$, $x = gyg^{-1}$ for some $y \in G_\rho(d)$. Hence $\rho(x) = \rho(g)\rho(y)\rho(g^{-1})d^* = d^*$ and $x \in G_\rho(d)$. Hence $gG_\rho(d)g^{-1} \subseteq G_\rho(d^*)$ and, since both groups have the same cardinality, the result follows.

<u>Appendix 2</u> (see page 139)

<u>Proof</u> Write $H = G_\rho(d_i)$. Let $D_i = \{\alpha_1, \alpha_2, \ldots, \alpha_s\}$, where $d_i = \alpha_1$. Then, since D_i is an orbit of ρ, there exist $g_j \in G$ such that

$$\rho(g_j)\alpha_1 = \alpha_j \quad (j = 1,2,\ldots,s) .$$

Consider the set of cosets $\{Hg_j : j = 1,2,\ldots,s\}$ of H in G. We have $Hg_j = Hg_k$ if and only if $g_j g_k^{-1} \in H$ which implies $j = k$. Clearly $G \subseteq \{Hg_j : j = 1,2,\ldots,s\}$. Hence $|G| = s|H|$.

<u>Appendix 3</u> (see page 139)

<u>Proof</u> Write $H = G_\theta(s)$. Then H is conjugate to G_j, for some j, $1 \leqslant j \leqslant N$. Let $S = \{\alpha_1, \alpha_2, \ldots, \alpha_t\}$, where $\alpha_1 = s$. Since S is a θ-orbit, for each t, $1 \leqslant i \leqslant t$, there exists $g_i \in G$ such that $\theta(g_i)\alpha_1 = \alpha_i$. Again, since S is an orbit, $G = \{Hg_i : i = 1,2,\ldots,t\}$ and the elements g_i are distinct. Furthermore $\rho(g)(Hg_i) = Hg_j$ if and only if $\theta(g)\alpha_i = \alpha_j$. Hence ρ and θ are equivalent.

Acknowledgements

I am indebted to the works of de Bruijn [2], [3], [4], [5], and de Bruijn and Klarner [6] whose presentation has considerably influenced this work.

Finally I would like to pay my deep respects to the families of J. Howard Redfield and my supervisor H.O. Foulkes.

REFERENCES

1. Burnside, W. (1911). Theory of groups of finite order. 2nd ed., C.U.P., Cambridge, reprinted Dover, New York, 1955.

2. de Bruijn, N.G. (1959). Generalization of Pólya's fundamental theorem in enumerative combinatorial analysis. Konink1. Ned. Akad. van Wetenschappen, 62, 59-69.

3. de Bruijn, N.G. (1964). Pólya's theory of counting, in Applied combinatorial mathematics, ed. E.F. Beckenbach. Wiley, New York.

4. de Bruijn, N.G. (1971). A survey of generalizations of Pólya's enumeration theorem. Nieuw Archief Wisk.(2), 19, 89-112.

5. de Bruijn, N.G. (1970). Recent developments in enumeration theory. Proc. Inter. Congress Math., Nice, 3, 193-199.

6. de Bruijn, N.G. and Klarner, D.A. Pattern enumeration, (unpublished manuscript).

7. Foulkes, H.O. (1963). On Redfield's group reduction functions. Canad. J. Math., 15, 272-284.

8. Foulkes, H.O. (1968). On Redfield's range-correspondences. Canad. J. Math., 18, 1060-1071.

9. Hall, M.,Jr. (1959). Theory of groups. Macmillan, New York.

10. Harary, F. (1960). Unsolved problems in the enumeration of graphs. Magyar Tud. Akad. Mat. Kutató Int. Közl., 5, 63-95.

11. Harary, F. & Palmer, E.M. (1973). Graphical enumeration. Academic Press, New York.

12. Harary, F. & Robinson, R.W. (1983). The rediscovery of J. Howard Redfield's papers. J. Graph Theory (to appear).

13. Littlewood, D.E. (1950). The theory of group characters and matrix representations of groups, 2nd ed., Clarendon Press, Oxford.

14. Neumann, P.M., (1979). A lemma that is not Burnside's. Math. Scientist, 4, 133-141.

15. Pólya, G. (1937). Kombinatorische Anzahlbestimmungen für Gruppen, Graphen und chemische Verbindungen. Acta Math., 68, 145-254.

16. Read, R.C. (1959). The enumeration of locally restricted graphs. J. London Math. Soc., 34, 417-436.

17. Redfield, J.H. (1927). The theory of group reduced distributions. Amer. J. Math., 49, 433-455. [R1]

18. Redfield, J.H. (1940). Enumeration by frame group and range groups. J. Graph Theory (to appear). [R2]

19. Robinson, R.W. (1970). Enumeration of non-separable graphs. J. Combin. Theory, 9, 327-356.

20. Sheehan, J. (1965). The superposition of graphs. Ph.D. thesis, University of Wales (Swansea).

21. Sheehan, J. (1967). On Pólya's theorem. Canad. J. Math., 19, 792-799.

22. Sheehan, J. (1968). The number of graphs with a given automorphism group. Canad. J. Math., 20, 1068-1076.

23. Stockmeyer, P.K. (1971). Enumeration of graphs with prescribed automorphism group. Ph.D. thesis, University of Michigan, Ann Arbor, Mich.

24. White, D.E. (1975). Counting patterns with a given automorphism group. Proc. Amer. Math. Soc., 47, 41-44.

25. Williamson, S.G. (1973). Isomorph rejection and a theorem of de Bruijn. SIAM J. Comput., 2, 44-59.

26. Wright, E.M. (1981). Burnside's Lemma: A historical note. J. Combin. Theory Ser. B, 89-90.

CHARACTERIZATIONS OF THE LIE INCIDENCE GEOMETRIES

Ernest Shult
Department of Mathematics
Cardwell Hall
Kansas State University
Manhattan, Kansas 66506
U.S.A.

1 INTRODUCTION

A very famous theorem (associated with the names Hilbert, von Staudt, Veblen and Young) characterizes projective spaces of dimension greater than 2 as linear incidence systems satisfying a certain (variously named) axiom. By the term "characterization", one means a complete classification in terms of division rings. This famous characterization theorem fully displays the spirit of synthetic geometry in that one obtains an exact and elaborate structure with many subspaces from a few simple axioms mentioning only points and lines.

More than three decades later F. Buekenhout and the author obtained a characterization of polar spaces of rank more than 2 in terms of a similar set of very simple axioms concerning only points and lines. But this time the characterization rested on a considerably more involved theory of Veldkamp and Tits, where, in effect, the really difficult work was done. Indeed Tits' work on polar spaces (as axiomatized by him) was a part of his monumental classification of buildings of spherical type of rank greater than 2. The Buekenhout-Shult polar space theorem could then be seen as a characterization of the buildings of types C_n and D_n in terms of axioms involving only two types of varieties of the building. The question was then raised (see [20]) whether similar axiomatically simple "point-line" characterizations could be obtained for all the buildings of spherical type of rank at least 3.

I am here to report that this goal has recently been achieved, thus fulfilling in some sense what the Veblen-Young theory pointed to all along: a full characterization of all the geometries which have a right to be called the "classical" geometries, purely synthetically.

To be sure, such a theorem involves results of many researchers patched together, but one feels bound to mention that by far the biggest gaps were filled by recent work of Arjeh Cohen and Bruce Cooperstein, both separately and jointly. As might be expected, lurking behind this work is Tits' basic classification of spherical buildings, but this time in the form of a far more amenable implement, his characterizations of diagram geometries of spherical type.

It appears then, that any reasonable description of the point-line characterization theorems of the title must incorporate three objectives: (1) to describe the geometries being characterized - that is the buildings and the Lie incidence systems obtained from them, (2) to describe the mechanics of the diagram geometries and Tits' characterizations of them, and finally (3) to describe how, from axioms on points and lines, one goes about building up hierarchies of subspaces which are to become the varieties of a fully reconstructed diagram geometry.

2 GRAPH-THEORETIC BACKGROUND

Let $\Gamma = (V,E)$ be a graph with vertex set V and edge set E. The edges are undirected; there are no loops. All subgraphs considered are induced subgraphs, and so we may unambiguously denote subgraphs by indicating the subset of V representing their vertex set. For any vertex v, the set of vertices adjacent to v is denoted $\Gamma(v)$ and the set $\{v\} \cup \Gamma(v)$ is denoted $v^{\perp}$. The intersection of sets $x^{\perp}$ as x ranges over X is written $X^{\perp}$; $\Gamma(X)$ denotes $X^{\perp} - X$.

A complete subgraph is called a <u>clique</u>; a subgraph of Γ which has no edges is a <u>coclique</u>. A graph Γ is said to be <u>multipartite with components</u> $V_1, V_2, \ldots$ if its vertex set V can be partitioned into subgraphs V_i which are cocliques.

A <u>path from</u> u <u>to</u> v is a finite sequence of vertices $(u = x_0, x_1, \ldots, x_n = v)$ with x_i adjacent to x_{i+1}, $i = 0, \ldots, n-1$, and the integer n is called its <u>length</u>. A path from u to v of minimal length d is termed a <u>geodesic</u>, and the integer $d = d(u,v)$ is the <u>distance</u> from u to v. The set of vertices at distance d from a fixed vertex v is denoted $\Gamma_d(v)$, so $\Gamma(v) = \Gamma_1(v)$ in our previous notation.

A path $(x_0, x_1, \ldots, x_n)$ is <u>contained</u> in a subgraph X if each x_i belongs to X. A subgraph X is called k-<u>convex</u> if every geodesic of length k between any two vertices of X lies in X. Clearly if X contains no pair of vertices at distance k, X is vacuously k-convex. Moreover, the intersection of any collection of k-convex subgraphs is k-convex and so any subgraph X of Γ lies in a unique smallest k-convex subgraph, the k-convex closure of X in Γ. A subgraph is always 1-convex, and is simply called convex if it is k-convex for all positive integers k. Again the intersection over any family of convex subgraphs is convex and so the convex closure of any subgraph X is well-defined.

A homomorphism $\Gamma \to \Gamma'$ between graphs is a mapping of the vertex set V of Γ into the vertex set V' of Γ' so that the image of any edge of Γ is either a single vertex or an edge of Γ'. If the mapping is 1-1 and onto, it is an isomorphism or an automorphism if it is an isomorphism from Γ to itself. Γ is called a cover of Γ' with respect to the homomorphism $f:\Gamma \to \Gamma'$ if f is surjective and induces by restriction an isomorphism $\Gamma(v) \to \Gamma'(f(v))$ for each vertex v. Let A be a group of automorphisms of $\Gamma = (V,E)$. We define the graph Γ/A by letting its vertex set be the set V/A of A-orbits on V, and asserting that the two A-orbits x^A and y^A are adjacent in Γ/A if they are distinct and at least one vertex of x^A is adjacent to a vertex of y^A. There is a canonical graph homomorphism $f_A:\Gamma \to \Gamma/A$ which sends each vertex x of Γ to its A-orbit x^A. It is easy to observe:

Proposition: Γ is a cover of Γ/A with respect to the canonical homomorphism f_A if and only if for each $(x,a) \in V \times A$, either $x = x^a$ or $d(x,x^a) \geq 4$.

Given a graph $\Gamma = (V,E)$, it is easy to construct covers in the following way. Let E' denote the set of ordered pairs (x,y) where $\{x,y\}$ is an edge in E, and let G be a group acting on a set M. A 2-cocycle is a mapping $\phi:E' \to G$ satisfying

$$\begin{aligned} &\phi(x,y) = \phi(y,x)^{-1}, \text{ for every } (x,y) \in E' \text{ and} \\ &\phi(x,y)\phi(y,z)\phi(z,x) = 1 \text{ for every 3-clique } \{x,y,z\}. \end{aligned} \tag{2.1}$$

The cover $\hat{\Gamma} = (\Gamma,\phi,G,M)$ has the cartesian product $V \times M$ as its vertex set, and two of its elements (x,m) and (y,n) are declared to be adjacent if and only if x is adjacent to y in Γ and

$$n = m^{\phi(x,y)} .$$

Then, because of (1.1), $\hat{\Gamma}$ is a cover of Γ with respect to the projection $p:\Gamma \times M \to \Gamma$.

Any mapping $\nu:V \to G$ is called a 1-cochain, and if $\phi:E' \to G$ is a 2-cocycle, then so also is the mapping $\phi^{\nu}:E' \to G$ defined by

$$\phi^{\nu}(x,y) = \nu(x)^{-1}\phi(x,y)\nu(y) \quad \text{for each} \quad (x,y) \in E'. \tag{2.2}$$

If $\hat{\Gamma}^{\nu} = (\Gamma,\phi^{\nu},G,M)$ is the cover of Γ obtained by using the cocycle ϕ^{ν} in place of ϕ, then the mapping $\hat{\Gamma} \to \hat{\Gamma}^{\nu}$ defined by sending each (x,m) in $\Gamma \times M$ to $(x,m^{\nu(x)})$ is an isomorphism $\hat{\nu}$ of covering graphs, which commutes with the projection maps.

Proposition 2.2. If $\hat{\Gamma}$ is a cover of Γ, then there exists a group G, a set M acted on by G, and a 2-cocycle ϕ such that $\hat{\Gamma} = (\Gamma,\phi,G,M)$. That is, every cover $\hat{\Gamma}$ arises by the construction above.

Now let $\Gamma = (V,E)$ be a connected graph and $\hat{\Gamma}$ a cover of Γ. Then $\hat{\Gamma}$ is said to be a trivial cover of Γ with respect to the homomorphism f if every connected component of $\hat{\Gamma}$ is isomorphic to Γ via f. From Proposition 2.2 one may deduce (on cohomological grounds) the following proposition, (which should be compared with Ronan [16]).

Proposition 2.3. Let Γ be a connected graph having a vertex v such that the following hold:

(i) $d(v,w)$ is bounded as w ranges over Γ.

(ii) If $w \in \Gamma_d(v)$, then the subgraph $w^{\perp} \cap \Gamma_{d-1}(v)$ is connected and contains at least two vertices.

(iii) If w_1 and w_2 are adjacent vertices of $\Gamma_d(v)$, then $w_1^{\perp} \cap w_2^{\perp} \cap \Gamma_{d-1}(v)$ is non-empty.

Then every cover of Γ is trivial.

In the next section we shall meet multipartite graphs Γ which are Tits geometries. For these graphs, something a little weaker than being a cover is required in order to obtain a graph $\hat{\Gamma}$ mapping homomorphically onto Γ so that $\hat{\Gamma}$ will belong to the same diagram as Γ. This notion requires that there be a certain collection S of cliques

$\hat{X}$ in $\hat{\Gamma}$ so that if $\hat{X}$ maps onto the clique X in Γ, then the morphism induces an isomorphism $\hat{\Gamma}(\hat{X}) \to \Gamma(X)$. (See the notion of *S*-cover in Ronan [16].)

3 DIAGRAM GEOMETRIES, BUILDINGS

A geometry is a multipartite graph Γ whose components V_σ are indexed by a set T whose members are called types. The type function $t:V \to T$ maps each vertex to the index of the component to which it belongs, so that if $v \in V_\sigma$, then $t(v) = \sigma$ and v is said to be a vertex of type σ. Since T is an index set t is an epimorphism. The cardinality of T is called the rank of Γ.

A clique in Γ is called a flag (this may include the empty clique). Two flags F_1 and F_2 are said to be incident if $F_1 \cup F_2$ is a flag. The type of a flag F is just the set $t(F)$ whose cardinality is called the rank of the flag. The set $T - t(F)$ is dually called the cotype of F and its cardinality the corank of F.

If F is a flag in Γ, the subgraph $\mathrm{Res}(F) = F^\perp - F$ is also multipartite and inherits a distinguished partition of its vertices into cocliques, namely

$$\{\mathrm{Res}(F) \cap V_\sigma | \sigma \in T - t(F),\ F^\perp \cap V_\sigma \neq \phi\}$$

and so is a geometry, the residual of F with type function t_F.

A flag F of the geometry Γ is called a chamber if $F \cap V_\sigma$ is non-empty for each type σ in T. The geometry Γ has the transversal property if and only if every clique lies in a chamber. This is equivalent (via the maximum principle) to the property:

For every flag F and type σ, F is incident with at least one vertex of type σ. (TP)

If Γ has the transversal property, then the residual geometry $\mathrm{Res}(F)$ possesses (non-empty) components $\mathrm{Res}(F) \cap V_\sigma$ for each type σ in $T - t(F)$ and so $\mathrm{Res}(F)$ has rank equal to the corank of F. Moreover $\mathrm{Res}(F)$ must also have the transversal property.

A geometry Γ is said to be residually connected if (1) every flag F of corank 1 lies in a chamber and (2) for each flag F

of corank 2, the subgraph $F^{\perp} - F = \mathrm{Res}(F)$ is connected and non-empty. Of course if Γ already has the transversal property, Γ is residually connected if and only if $\mathrm{Res}(F)$ is connected for each flag F of corank 2. If Γ has rank at least 2, then residual connectedness implies Γ is a connected graph since $\mathrm{Res}(F) = \Gamma$ when Γ is the empty flag. Note that residual connectedness inherits to residual geometries. All of the diagram geometries to be considered shortly are residually connected and have the transversal property.

Let Γ be a geometry and let V_σ be one of its components, so V_σ is the set of all vertices of type σ. For any flag F of Γ, the set $\mathrm{Sh}_\sigma(F) = F^{\perp} \cap V_\sigma$ is called the σ-<u>shadow</u> of F (note that $\sigma \in t(F)$ is allowed). The use of shadows allows us to "realize" the geometry Γ in terms of a "geometry" of subsets of V_σ. For example, we may have on the one hand an abstract incidence system with objects of various types and an incidence relation which can only come in force between objects of different types. On the other hand, we may wish to view one set of objects as "points" and, as is often done in geometry, view the remaining objects as special sets of points - despite the fact that this may force consideration of "repeated objects". It is this shift from the former view to the latter, that the process of forming shadows formalizes.

The language of shadows allows us to discuss further properties of geometries. The first of these is called by Tits, a "linearity condition" [25] (see also Buekenhout [2]):

> The system of σ-shadows of flags of Γ, together with the empty set, is closed under finite intersections. (GL)

A stronger condition is the following (see Buekenhout [2]):

> Let F be a flag and x a vertex of Γ. Then either F and x have disjoint shadows on V_σ or else there is a flag F' incident with both F and x such that $\mathrm{Sh}_\sigma(F) \cap \mathrm{Sh}_\sigma(x) = \mathrm{Sh}_\sigma(F')$. (Int)

Assume Γ is a geometry of finite rank and is connected. An m-<u>cover</u> of Γ is a type preserving surjection $f: \hat{\Gamma} \to \Gamma$ of geometries, which by restriction induces an isomorphism $Res(\hat{X}) \to Res(X)$ for any flag $\hat{X}$ of corank m in $\hat{\Gamma}$. (If rank $\Gamma = n$, an $(n-1)$-cover is just a cover in the sense of section 2, with respect to a morphism preserving types). In [25] Tits has shown that a geometry Γ satisfying the hypothesis of this paragraph possesses a universal m-cover $u:\Gamma_0 \to \Gamma$, that is one for which the morphism u may be factored through any other m-cover $f:\hat{\Gamma} \to \Gamma$ (see also Aschbacher [1] for an approach to this result using presheaves).

We may now turn our attention to diagram geometries. The use of the concept of diagram geometry is to impose on a geometry Γ a uniformizing assumption on the residue geometries of Γ. In particular one wishes to impose the assumption that whenever F is a flag of corank 2 in Γ, then Res(F) belongs to a category of rank 2 geometries which depends only on the type of F. To facilitate the discussion, we assume to the end of this section that all geometries Γ have finite rank, have the transversal property and are residually connected, all three of these properties being preserved by morphisms and the taking of residuals.

A rank 2 geometry Γ is just a bipartite graph and as such may be viewed as the incidence graph of an incidence system (P,L) of "points" and "lines". The incidence system (P,L) is connected and every point is incident with at least one line and vice versa from the assumptions on Γ of the preceding paragraph. The diagram geometries are built up from a glossary of rank 2 geometries, some of whose most useful members we now describe.

A digon is an incidence system (P,L) in which every point is incident with every line. (Note that since P and L are components of a partition of the vertices of Γ, they are non-empty sets.) (P,L) is a linear incidence system if and only if $|P| > 1$ and two points of P lie on at most one line of L. (This concept is called variously "partial linear space" or "semilinear space" elsewhere [10], [11], [22].) Since this simply means Γ has no subgraphs isomorphic to $K_{2,2}$ it is self-dual, that is, (L,P) is also a linear incidence system. A linear space is a linear incidence system (P,L) in which every pair of points lies on a line. If also (L,P) is a linear space, then (P,L) is the familiar projective plane (noting that lines with two points are allowed). A rank 2 geometry Γ of diameter m and girth 2m is called a <u>generalized</u> m-<u>gon</u>. It follows that a generalized 2-gon is a digon with at least 3 vertices and a generalized 3-gon is a projective plane.

We wish to attach a diagram consisting of two nodes and an edge labeled K to any category K of rank 2 geometries. The left to right convention of reading the "K" imposes a left to right orientation of the two nodes and with this convention the left hand node represents the set of points P and the right hand node the lines L. Thus to express the dual incidence systems one must write the K backwards. For many special rank 2 geometries the label "K" is replaced by special symbols, for example "Af" for affine planes, "C" for thin projective spaces,etc. For generalized m-gons the edge is labeled (m) and for small values of m it is conventional merely to replace the edge by a multiple bond of m - 2 strands. Thus a digon is represented by the diagram consisting of two disconnected nodes, a projective plane by two nodes connected by a single bond etc.

A <u>diagram</u> Δ is a finite graph whose edges are labeled with one of the conventional symbols (m), Af, C, etc., or in general by symbols K denoting various categories of rank 2 geometries. A <u>diagram geometry</u> <u>Γ with diagram</u> Δ is a geometry Γ and a diagram Δ whose nodes are the set T of types of Γ with the inductive property:

> For any flag F of type t(F), the residual geometry Res(F) is a diagram geometry with diagram Δ - t(F). (D)

A <u>geometry of type</u> M is a diagram geometry all of whose rank 2 residuals are generalized m-gons. For example, if Δ is the diagram O═O═O, then a diagram geometry Γ with diagram Δ is a rank three geometry whose types may be regarded as points, lines, and generalized quadrangles, reading nodes from left to right. Any point incident with a line incident with a quadrangle is also incident with the quadrangle, a fact discovered by observing that suppression of the central node leaves a digon diagram. The system of lines and quadrangles incident with a fixed point p themselves form the incidence system of a generalized quadrangle (by suppression of the left node). Thus we see that the diagram is a convenient way of imposing uniform axioms on a geometry Γ.

There is a special collection of diagrams called *the spherical diagrams*. They include the diagrams $\circ\overset{(m)}{\text{———}}\circ$ for the generalized m-gons. The remaining diagrams have three or more nodes and are listed in Table 1. Since we shall need to make special reference to various nodes in describing the Lie incidence systems, we impose a numbering system on the nodes of the spherical diagrams of Table 1.

A geometry Γ is called *thin* if every flag of corank 1 lies in exactly two chambers. A thin diagram geometry Γ whose diagram is a diagram of spherical type is called a *Coxeter geometry*. A Coxeter geometry of rank 2 is thus just an ordinary m-gon. Each Coxeter geometry is uniquely determined by its diagram. For example, a Coxeter geometry with diagram A_n has as its vertices of type k, the k-subsets of a fixed $(n + 1)$-set, $k = 1,\dots,n$, with incidence being containment.

A subgraph A of a geometry Γ of type M is called an *apartment* if it is a Coxeter geometry with respect to the system of types $\{A \cap V_\sigma\}$ inherited from Γ. Thus A and Γ are geometries of the same rank. A geometry Γ of type M, is said to be a *building of type* M if Γ possesses a system $\mathcal{A}$ of apartments such that the following hold:

Every pair of flags in Γ lie in a common apartment in $\mathcal{A}$. (Ap 1)

If two apartments A_1 and A_2 in $\mathcal{A}$ contain two flags F_1 and F_2 in common, then there is a type-preserving isomorphism $A_1 \to A_2$ fixing $F_1 \cup F_2$ vertex-wise. (Ap 2)

We may now state Tits' great classification theorem [24]:

Theorem 3.1 Every building of rank at least three of spherical type - that is every building whose diagram appears in Table 1 - is known.

Now a building Γ of spherical type is *simply 2-connected*, which means that any connected 2-cover of Γ is isomorphic with Γ. This means that if any geometry of type M has a building for a 2-cover then that building is a universal 2-cover. Not every universal

Table 1. The spherical diagrams of rank at least three.

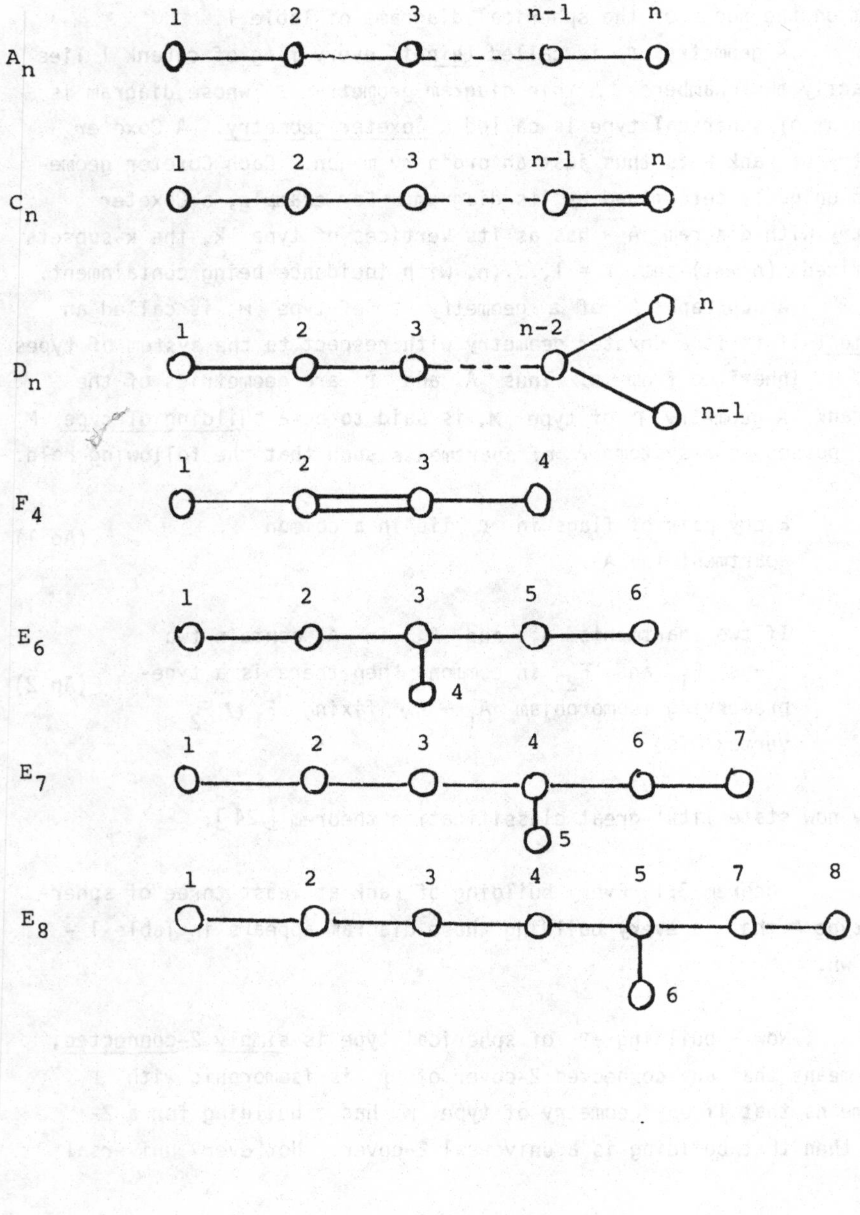

cover of a geometry of type M, or even one of spherical type is a building. But Tits' second theorem tells us that this case can be decided by looking at the residues which have diagrams of type C_3 and H_3, where H_3 denotes the diagram $\circ\!-\!\!-\!\circ\overset{(5)}{-\!\!-}\!\circ$. Specifically:

Theorem 3.2 (Tits [25]) If Γ is a geometry of type M and if all residues of type C_3 and H_3 are covered by buildings, then $\Gamma \simeq B/A$ where B is a building and A is a group of automorphisms of A where, for each flag X in B, the canonical mapping $\pi: B \to B/A \simeq \Gamma$ induces an isomorphism

$$\mathrm{Res}(X)/\mathrm{Stab}_A(X) \to \mathrm{Res}_{(B/A)}(\pi(X)).$$

The condition on A in the theorem is equivalent to asserting:

Every flag of $\Gamma = B/A$ is the image X^A of a flag X of B. (A1)

If $\{v_1,\ldots,v_n\} = X$ is a flag of B, then A transitively permutes the set of all flags of B lying in $v_1^A \cup v_2^A \cup \ldots \cup v_n^A$. (A2)

(Compare with conditions (Q1)' and (Q2)' of Tits [25] which work without finiteness of rank.) Several rich corollaries follow from this theorem, but one of the most useful for characterizing the Lie incidence systems is this one:

Theorem 3.3. Let Γ be a geometry of spherical type of rank at least three. If Γ satisfies (Int) then Γ is a building and hence is known.

4 POINTS AND LINES

Let (P, L) be an incidence system, with P the set of <u>points</u> and L the set of <u>lines</u>. If " " denotes the relation on P of being collinear, its graph $\Gamma = (P,\sim)$ is called the collinearity graph. Since the rank 2 geometry which goes with (P, L) is then the bipartite graph $\Gamma_I = (P \cup L, I)$ where I is the incidence relation between points and lines, the distance between points in Γ_I is twice that in Γ. If every line is incident with at least one point, Γ_I is connected if and only if Γ is.

For the rest of this paper, we assume (TP) for Γ_I, so that every point lies on at least one line and vice versa. With some abuse of language, we will often denote the point-shadow of a line L by the same symbol L. The graph-theoretic notation of section 2 will apply in full to the collinearity graph Γ as well as to Γ_I.

A subspace X is a subset X of P such that $|L \cap X| \geq 2$ implies $L \subseteq X$ for any line L. We say X is a singular subspace of (P,L) if X is a subspace and is a clique in Γ.

A line L is called thin if it is incident with only two points. We say (P,L) is the product of subspaces $X_1, X_2 \ldots$ if the X_i are subspaces which partition P and if, for i,j $(i \neq j)$ $(x_i, x_j) \in X_i \times X_j$ implies that $\{x_i, x_j\}$ is a thin line.

An incidence system (P,L) is called a gamma space (following Higman [14]) if for every point-line pair (p,L), $p^\perp \cap L$ is empty, consists of a single point or is all of L. This is equivalent to requiring, for every subset X of P, that X always be a subspace. A strong gamma space is an incidence system (P,L) which, for every point-line pair (p,L) either $p \in L$ or if d is the smallest integer for which $\Gamma_d(p) \cap L$ is non-empty, then $|\Gamma_d(p) \cap L| \geq 2$ implies $L \subseteq \Gamma_d(p)$. If $\Gamma^*_d(p)$ is the set of all points of P at distance less than or equal to d from p then (P,L) is a strong gamma space if and only if $\Gamma^*_d(p)$ is a subspace of (P,L) for every point-line pair p,L and non-negative integer d.

Now consider a building B with spherical diagram Δ. We fix a type σ, which of course corresponds to a node (also designated "σ") of Δ. We may then consider the geometry of shadows on V_σ. Thus if we view $V_\sigma = P$ as a set of points, which flag shadows $Sh_\sigma(F)$ should we use as a set of lines L so that (P,L) is a gamma space? This question was answered by Cooperstein [10]. Let $\Delta(\sigma)$ be the set of all nodes of Δ which are "adjacent" to σ in the diagram Δ. (Recall that no edge connects nodes representing a digon. All other rank 2 geometries are represented by "adjacent" nodes. This is the "basic diagram" in the sense of Buekenhout [2]). Then, setting $L = \{Sh_\sigma(F) | F \text{ a flag with } t(F) = \Delta(p)\}$, we have (P,L) a gamma space.

As a bonus it is a linear incidence system as well. The geometry (P,L) formed from a building B and a node σ in this way is called a Lie incidence system.

It should be clear from this that a Lie incidence system is determined by specifying a spherical diagram, and then one of its nodes, the latter being identified by the numbering displayed in Table 1. For example, $A_{5,2}$ (or $A_{5,2}(K)$, if one wants to indicate the relevant division ring) is the geometry whose points are the 2-subspaces of a 6-dimensional vector space V over a division ring K and whose lines are the (1-space, 3-space)-flags of V. A point is incident with a line, if together they form a (1-space, 2-space, 3-space)-flag of V. Such a space is an example of a Grassman space which we shall meet later on.

Certain species of linear incidence systems recur in many guises inside the Lie incidence systems and so need to be discussed.

A projective space is a linear space (P,L) satisfying the following axiom:

> If L_1, L_2 and L_3 are three lines intersecting at three distinct points, then any line L_4 intersecting L_1 and L_2 but not at their point of intersection, intersects L_3. (P)

In a projective space with no thin lines it is easy to show that all lines possess the same cardinality. We call these thick projective spaces. Any projective space is a product of thick projective spaces.

If (P,L) is a projective space, it is possible to show that the set of all subspaces is a semi-modular poset and so if there is an upper bound on the length of a chain of subspaces, a rank function is defined. The resulting geometry Γ whose vertices are the proper subspaces, whose type function is the rank function, is a diagram geometry with diagram A_n for some n. Since the converse also holds we have a 1-1 correspondence of diagram geometries of type A_n and projective spaces of bounded rank.

A polar space is an incidence system (P,L) satisfying:

> If $(p,L) \in P \times L$, then either $L \subseteq p^{\perp}$ or $|p^{\perp} \cap L| = 1$. (Po)

Clearly a polar space is a species of gamma space, but the assumption (Po) is rather "global" in that it implies that the collinearity graph Γ has diameter at most 2. The radical $\mathrm{rad}(P) = \{p \in P \mid P \subseteq p^{\perp}\}$ is a subspace, and if $\mathrm{rad}(P)$ is empty (P,L) is said to be non-degenerate. If (P,L) is a non-degenerate polar space, it is a linear incidence system [4]. For any polar space (P,L) there is a canonical procedure for "factoring out the radical" to obtain a non-degenerate polar space (or the empty set, if $\mathrm{rad}(P) = P$)[4], [15]. In this way one reduces to the consideration of non-degenerate polar spaces.

Let P be the projective space of some vector space V and P* the projective space of the dual vector space V*. For each $p \in P^*$ there is a hyperplane $p^{\perp} \cap P$ representing the kernel of all functionals represented by p. We form an incidence system (P_1, L_1) where $P_1 = P \cup P^*$ and where L_1 consists of all lines of the projective spaces P and P* together with the collection of thin lines {q,p} where p ranges over P* and $q \in P \cap p^{\perp}$. Then (P_1, L_1) is a non-degenerate polar space with thin lines which we say is of alternating type. If (P,L) is any non-degenerate polar space, then it is a product of thick polar spaces and polar spaces of alternating type. (Buekenhout and Sprague [5]).

Let (P,L) be a non-degenerate polar space. The collection S of all proper singular subspaces is a lower semi-lattice under inclusion. If S has an unrefinable chain of finite length r, then all unrefinable chains in S have length r [15] and r is called the rank of the polar space (P,L). ("Rank" can then be defined canonically for all polar spaces degenerate or not.)

A near n-gon is an incidence system (P,L) satisfying the following axioms:

The collinearity graph $\Gamma = (P,\sim)$ has finite diameter. (N1)

For each point p and line L, L contains a unique point nearest p. (N2)

"Nearest" is defined, of course, in the metric of the collinearity graph Γ. The axioms imply, as an aside, that (P,L) is a linear incidence system. Generalized n-gons are near n-gons, and if diam $\Gamma = 2$, a near n-gon is a generalized 4-gon. Accordingly we adopt the convention of setting the otherwise free parameter "n" equal to twice the diameter of Γ. Taking the 759 octads of the Steiner system $S(24,5,8)$ as points and mutually disjoint triples of octads as lines, one obtains a near hexagon which is not a generalized hexagon. A near n-gon $(\mathcal{P},\mathcal{L})$ is also a strong gamma space and the axioms are quite global in that (N2) makes reference to points and lines which may be very far apart in the incidence graph Γ_I.

A fundamental result for near n-gons $(\mathcal{P},\mathcal{L})$ is :

Yanushka's Lemma. If p and q are two points at distance 2 in Γ, and if there are at least two paths in Γ from p to q one of which involves a thick line, then the convex closure of $\{p,q\}$ in Γ is a subspace of $(\mathcal{P},\mathcal{L})$ which is a generalized quadrangle.

If for every pair of points at distance two in $(\mathcal{P},\mathcal{L})$, the convex closure is a generalized quadrangle (as a subspace of $(\mathcal{P},\mathcal{L})$) we say that <u>quads exist for</u> $(\mathcal{P},\mathcal{L})$ and call the convex subspaces themselves <u>quads</u>. If p is a point and Q is a quad not containing p and Q is not thin, then the set of points in Q nearest p is either a single point (the <u>classical</u> relation) or is an ovoid in Q (the ovoid relation) [21]. A near n-gon is said to be <u>classical</u> if quads exist and every non-incident point-quad pair bears the classical relation. It is said to be <u>near classical</u> if the classical relation is imposed on p and Q only when p has distance 2 from its nearest point in Q (from convexity of Q classicalness is automatic if p has distance one from its nearest point in Q).

We now conclude this section by listing the basic characterization theorems for the linear incidence systems we have considered here.

Theorem 4.1 (von Staudt, Hilbert, Veblen-Young)

A thick projective space of rank $n \geq 3$ is the geometry of subspaces of an $(n + 1)$-dimensional vector space over some division ring.

Theorem 4.2 (Veldkamp, Tits, Buekenhout-Shult).

A thick non-degenerate polar space of rank ≥ 3 is the Lie incidence system $C_{n,1}$, $n \geq 3$, (and hence is known by Tits' Theorem of section 3).

Theorem 4.3 (Cameron [6]).

A classical near 2n-gon for $n \geq 3$, is the Lie incidence system $C_{n,n}$ and hence is known.

5 COOPERSTEIN'S THEORY

An account of this theory is given by Buekenhout in [3]. Like Yanushka's Lemma of section 4, it gives another kind of subspace to serve as a node in a diagram geometry, in addition to the two representing points and lines.

We assume (P,L) is an incidence system satisfying the following axiom:

> If p and q are two points of P at distance two in the collinearity graph $\Gamma = (P,\sim)$ then either $p^{\perp} \cap q^{\perp}$ consists of a single point or else $p^{\perp} \cap q^{\perp}$ is a non-degenerate polar space of finite rank at least two. (F3)

If in (F3) we were to require $p^{\perp} \cap q^{\perp}$ to be a subspace of (P,L), it would easily follow from this slightly stronger axiom that (P,L) is a gamma space. Since (F3) usually accompanies the gamma space assumption anyway, there is not much need for such an "improvement". If p and q are as in (F3), we call $\{p,q\}$ a <u>special pair</u> if $|p^{\perp} \cap q^{\perp}| = 1$ and we call $\{p,q\}$ a <u>polar pair</u> (following Buekenhout [3]) if $p^{\perp} \cap q^{\perp}$ is a non-degenerate polar space. The main result of Cooperstein's theory is this:

Theorem 5.1.

(1) If (P,L) is a gamma space satisfying (F3), then for every polar pair $\{p,q\}$, its 2-convex closure is a subspace S(p,q) which is a non-degenerate polar space of rank one more than the rank of $p^{\perp} \cap q^{\perp}$.

(2) If in addition, it is assumed that for any pair of collinear points a,b, there exists a point c collinear with b but forming a polar pair with a, then (P, L) is a linear incidence system.

(3) Under the stronger assumption that for every 3-clique {a,b,c} in Γ there is a point d in $a^{\perp} \cap b^{\perp}$ forming a polar pair with c, then every singular subspace of (P, L) is a projective space.

The subspaces S(p,q) appearing in the theorem are called symps. The hypotheses of (2) and (3) just assert that every line or every plane lies in some symp. The original version of this theorem was proved by Cooperstein in [11] in a context in which geometries were finite, all lines were thick and no special pairs were present. Finiteness was never used in his proof, and Kantor observed soon after that special pairs could be allowed. The presence of thin lines required only minor modifications in Cooperstein's original proof (see Buekenhout [3] and Proposition 3.8 of Cohen [7]).

The power of this theory lies in the fact that its applications may arise in unexpected circumstances. The next two sections give two applications of Cooperstein's theory which while a little novel, at the same time yield characterizations of Lie incidence systems.

6 THE TALLINI-SPRAGUE THEOREMS

Our first application of Cooperstein's theory is a direct generalization of a theorem of Tallini [23] which characterized the Grassman spaces $A_{n,2}$. The generalization presented here is equivalent to an infinite version of a theorem of Sprague [22]. (It is the writer's understanding that Buekenhout and Sprague have since adjusted Sprague's original proof to free it of these finiteness assumptions.)

Theorem 6.1. Assume (P, L) is a connected incidence system satisfying the following assumptions:

(T1) Any three mutually collinear points lie in a singular subspace.

(T2) The maximal singular subspaces of (P, L) fall into two families Σ and Π such that

(i) any member of Σ meets any member of Π at a line or the empty set, and

(ii) every line lies in exactly one member of Π and exactly one member of $\sum$.

(T3) (An intersection property) If two distinct members P_1 and P_2 of Π each intersect non-trivially two members S_1, S_2 of $\sum$, then $P_1 \cap P_2$ is non-empty.

(T4) (Finite rank). Some member P of Π has an unrefinable chain of non-empty subspaces of length $d + 1$.

Then if L_0 is the set of lines which are actually intersections of members of Π with members of $\sum$, then (P, L_0) is the Grassmann space of d-subspaces of some vector space, and if $\sum$ has finite rank, it is a Lie incidence system of type $A_{n,d}, d > 1$.

One notes that the assumption that (P,L) is a linear incidence system is missing in (T1)-(T4). (P,L) is not itself a linear space since there are two classes of maximal singular subspaces, and so now it is clear that the collection of lines L^* which do not lie properly in any other line coincides with the set of lines L_0 of the theorem and that (1) (P,L_0) is a linear incidence system (because of (T2)) (2) L-collinearity is coextensive with L_0-collinearity on $P^{(2)}$ and (3) (P,L_0) is a gamma space.

Let Γ be the collinearity graph of (P,L_0). We can deduce (F3) from our axioms (actually so that no special pairs occur). Let $d(p,q) = 2$ for two points p and q in Γ. Then there is a point r in $p^\perp \cap q^\perp$, and singular spaces P_{pr} of Π on the line pr and S_{rq} in $\sum$ on the line rq. Then $P_{pr} \cap S_{rq}$ is a line in $p^\perp \cap q^\perp$ on r. In particular $\{p,q\}$ is not a special pair.

Now assume L is a line (of L_0) in $p^\perp \cap q^\perp$. By symmetry of p and q we may assume $\langle p,L\rangle \subseteq S_p \in \sum$ and $\langle q,L\rangle \subseteq P_q \in \Pi$. Suppose $z \in p^\perp \cap q^\perp - L$. Then if P_p is the element of Π on pz and S_q is the element of $\sum$ on qz, we see

$$P_p \cap S_q, \; P_p \cap S_p, \; P_q \cap S_q, \; P_q \cap S_p$$

are all lines of L_0. Then by (T3), $P_p \cap P_q = \{w\}$. Then $\{x\} \cup L \cup \{w\}$ is a clique and by (T2) lies in S_p. This places w in $L = S_p \cap P_q$ and so $z^\perp \cap L$ contains w. Thus $p^\perp \cap q^\perp$ is a polar space. Non-degeneracy follows also from (T2) and so the axiom (F3) holds. Thus by Theorem 4.1 all singular spaces in Σ and Π are projective spaces, and the 2-convex closure of every pair of points at distance 2 is a symp, which by (T3) is a polar space of type $D_{3,1}$. We denote the system of all symps S.

By assumption (T4), some projective space in Π has rank d. If two members P_1, P_2 of Π meet at a point p, there is a 1-1 correspondence between the lines on p lying in P_1 and the lines on p lying in P_2, with corresponding lines lying in a Σ-space. By forming symps containing corresponding pairs of lines, this correspondence takes planes of P_1 on p to planes of P_2 and so P_1 and P_2 are isomorphic projective spaces. Similar assertions hold with the roles of Σ and Π reversed. Let $(\Sigma,\sim)$ and $(\Pi,\sim)$ denote the graphs on Σ and Π with adjacency being intersection at a point. Connectedness of (P,L_0) implies $(\Sigma,\sim)$ and $(\Pi,\sim)$ are both connected and so all spaces in Π have rank d and all spaces in Σ are isomorphic.

Let T_1 and T_2 be two members of Σ meeting at a point p. Let $S(T_1,T_2) = \{S \in S \mid S \cap T_i$ is a plane, $i = 1,2\}$. Set $C(T_1,T_2) = \{T \in \Sigma \mid T \cap S$ is a plane for some $S \in S(T_1,T_2)\}$. One can then prove that $C(T_1,T_2)$ is a maximal clique in $(\Sigma,\sim)$. Cliques of this form in $(\Sigma,\sim)$ are denoted C. We have of course, another class of maximal cliques in $(\Sigma,\sim)$, namely those of the form $\Sigma_p = \{T \in \Sigma \mid p \in T\}$. This class can unambiguously be denoted P. Let $L(\Sigma)$ be the class of all subsets of Σ of the form $\Sigma_p \cap C$ where $C \in C$ and p is the intersection of at least two members of C. Then it is easy to verify that the incidence system $(\Sigma, L(\Sigma))$ also satisfies the axioms (T1)-(T4) with C and P playing the roles formerly occupied by Σ and Π. But now all members of P are projective spaces in $(\Sigma,L(\Sigma))$ of rank d-1 so induction on d may be applied to identify $(\Sigma,L(\Sigma))$ uniquely. The rest is a matter of manufacturing the isomorphism of (P,L_0) with the appropriate Grassman space from that now existing for $(\Sigma,L(\Sigma))$. (In case rank $T = 2d + 1 = n$ for all $T \in \Sigma$, there are two ways to do this.)

7 NEAR CLASSICAL NEAR n-GONS

Let (P,L) be a near n-gon with quads, that is each pair of points at distance two has a 2-convex closure which is a generalized quadrangle. We let $\mathcal{Q}$ denote the system of quads. If $Q \in \mathcal{Q}$ and p is a point not in Q, then $p^{\perp} \cap Q$ contains at most one point since (P,L)

has no triangles and Q is 2-convex. Recall that (P,L) was called classical if whenever $d = d(p,Q) = \min(d(p,q)|q \in Q), d \geq 1, \Gamma_d(Q)$ consists of a single point, and was called near classical if the conclusion $|\Gamma_d(p) \cap Q| = 1$ is asserted only for $d = 1$ or 2. Cameron proved that classical near n-gons were dual polar spaces (Theorem 4.3), but it has been thought for a long time that a conclusion similar to this should hold for the case that (P,L) is merely near-classical.

As a second application of Cooperstein's theory we will show that this is roughly the case, though this result is almost certainly contained in some guise in recent work of Ronan on truncated buildings ([17] [18]).

We assume then that (P,L) is a near classical near n-gon. We make first two observations which do not depend on the near-classical property: (1) any 5-circuit $\{x_1,x_2,\ldots,x_5\}$ lies in some quad and (2) if three quads Q_1,Q_2,Q_3 meet pairwise at lines, then all three meet at a point.

Let $L \in L$ and let Q_1 and Q_2 be quads meeting L at distinct points p_1 and p_2 respectively. Let X_i be the set of quads on L meeting Q_i at a line (necessarily on p_i), $i = 1,2$. Each line of Q_i on p_i matches a quad of X_i. What if $X_1 \cap X_2$ contains two quads R_1 and R_2? Let $L_{ij} = R_i \cap Q_j$. If w is any element of $Q_1 - p_1^{\perp}$, $w^{\perp} \cap L_{i1} = r_{i1}$, $i = 1,2$. Then w is distance two from the two distinct points $r_{11}^{\perp} \cap L_{12} = r_{12}$ and $r_{21}^{\perp} \cap L_{22} = r_{22}$ of Q_2 and so, from the near classical hypothesis, w is collinear with a unique point $f(w)$ of Q_1 in $r_{12}^{\perp} \cap r_{22}^{\perp}$. Thus on the line $L' = wf(w)$ there are two quads R_1' and R_2', being the convex closures of the respective 4-circuits $(w,f(w),r_{21},r_{11})$ and $(w,f(w),r_{22},r_{12})$, meeting Q_1 and Q_2 at lines. Thus repeating the above arguments, with (L',R_1',R_2') in the roles of (L,R_1,R_2) we see that any point of $Q_1 - w^{\perp}$ is collinear with a unique element of Q_2. Since w may range freely over $Q_1 - p_1^{\perp}$ in the previous statement we see that there is a 1-1 correspondence $f:Q_1 \to Q_2$ given by setting $f(x)$ to be the unique element of Q_2 collinear with x in Q_1. It is easy to see that this correspondence must preserve collinearity in Q_1 (so f is an isomorphism of quadrangles). In this case we call the two quads Q_1 and Q_2 an opposite polar pair. We have shown so far, that if two quads which do not intersect at a point intersect two further quads R_1 and R_2

at lines and if $R_1 \cap R_2$ is itself a line meeting both quads, then the original pair of quads is an opposite polar pair.

Next assume the disjoint quads Q_1 and Q_2 meet quads R_1 and R_2 at lines $L_{ij} = R_i \cap Q_j$ and that the lines L_{1j}, L_{2j} are not assumed to intersect in Q_j. Choosing point a on L_{11} in Q_1, $a^{\perp} \cap L_{12} = b$ and $a^{\perp} \cap L_{21} = c$ are themselves collinear with points $d = b^{\perp} \cap L_{22}$ and $d' = c^{\perp} \cap L_{22}$. Then d and d' are two points of $\Gamma_2(a) \cap Q_2$ and the near classical hypothesis forces $d = d'$. Then if the quad R is the convex closure of the 4-circuit (abdc) then R and R_1 satisfy the hypotheses of the previous paragraph, and so Q_1 and Q_2 are an opposite polar pair.

Finally, assume $Q_1 \cap Q_2 = \{p\}$, $Q_i \in \mathcal{Q}$, $p \in P$. Then for each line L_i on p in Q_i, the pair L_1, L_2 generates a quad R and conversely any quad R meeting both Q_1 and Q_2 at lines must contain p by (1). In this case the collection (Q_1,Q_2) of quads meeting both Q_1 and Q_2 at lines has the structure of a grid.

The upshot of the previous three paragraphs is this: Form a graph $\bar{\Gamma} = (\mathcal{Q},\sim)$ whose vertex set is $\mathcal{Q}$, two quads being declared adjacent if and only if they meet at a line. If we form lines $L(\mathcal{Q})$ in $\bar{\Gamma}$ as sets $\{x,y\}^{\perp\perp}$ where x and y are adjacent vertices in the graph $\bar{\Gamma}$, then a "line" is a set of quads $\{R_\sigma\}$ lying on a line L, all of which meet a quad Q at a line, where $Q \cap L$ is a point. By the third paragraph above, $(\mathcal{Q},L(\mathcal{Q}))$ is a linear incidence system (a hypothesis which appears under a different title in Ronan's first paper on truncated buildings), by (2) is a gamma space and by the three preceding paragraphs above, satisfies (F3). Thus $(\mathcal{Q},L(\mathcal{Q}))$ has symps.

Let H consist of only those symps which result when Q_1 and Q_2 are an opposite polar pair in $(\mathcal{Q},L(\mathcal{Q}))$. Then H is a class of 3-convex near hexagons covering every pair of points at distance 3. (The existence of 3-convex sub-near hexagons was proved in [19] by a direct method which required thick lines.)

This process can be duplicated up the line to get "higher" subspaces. Suppose now H_m is a class of sub-near m-gons. We assume (1) each member of H_m is a dual polar space of rank m and (2) if L is a line and $H \in H_{m-1}$ with $L \cap H$ being a point, then the convex closure of $L \cup H$ is an element of H_m. Then quite analogously with the arguments above, one may show that if $H_1, H_2 \in H_m$ and $R_1, R_2 \in H_{m-1}$

such that (i) $H_1 \cap H_2 = \emptyset$, (ii) $R_i \cap H_j = S_{ij} \in \mathcal{H}_{m-2}$, (iii) $R_1 \cap R_2 = S \in \mathcal{H}_{m-2}$, then (a) each point x of H_1 is collinear with a unique point x' of H_2 and (b) if x is collinear with y within H_1 then x' is collinear with y' in H_2. We call H_1 and H_2 an opposite polar pair in this case, define lines $\mathcal{L}(\mathcal{H}_m)$ from $(\mathcal{H}_m,\sim)$ where "$\sim$" indicates intersection at a member of $\mathcal{H}_{m-1}$. Then $(\mathcal{H}_m,\mathcal{L}(\mathcal{H}_m))$ again satisfies (F3) and is a gamma space. Cooperstein's theory once again gives us symps and we may set $\mathcal{H}_{m+1}$ to be those symps obtained from opposite polar pairs. The result is a geometry G with types $\mathcal{P},\mathcal{L},\mathcal{Q},\mathcal{H}_3,\mathcal{H}_4,\ldots,\mathcal{H}_{n-1}$ (where any member of $\mathcal{H}_n$ is $\mathcal{P}$) which is a diagram geometry of type C_n.

Although elements of $\mathcal{H}_3$ are 3-convex it is not clear from the construction that $\mathcal{H}_m$ is m-convex for $m > 3$. Thus the intersection property (Int) may well fail. None the less by Tits' theorem, G is B/A where B is a building of type C_n and $A \leq Aut(B)$ subject to the conditions (A1) and (A2) of section 3. Thus in some sense we know the geometry. But conversely, it is not clear when B/A is a near-classical near n-gon in the case when the images of the vertices of type n are taken as points.

8 COHEN'S CHARACTERIZATION OF $F_{4,1}$

This beautiful theorem of Cohen utilizes both of the theorems presented so far which guarentee new species of subspaces to produce "new" nodes of a diagram geometry, namely Yanushka's Lemma and Cooperstein's theory. Once all four nodes and a diagram geometry of type F_4 are obtained we have a geometry called a <u>metasymplectic space</u> by Freudenthal [13]. A metasymplectic space is a Tits geometry Γ of rank 4 consisting of four types of objects: points, lines, planes and symplecta (non-degenerate rank 3 polar spaces) with planes and symplecta being subspaces such that

(M1) The intersection of two distinct symplecta is empty, is a point, a line or a plane.

(M2) A symplecton with its incident points and lines is a non-degenerate polar space of rank 3.

(M3) Given a point p, the set of all lines L_p and planes Π_p containing p is a dual polar space of rank 3 (i.e. a point residual is a diagram geometry of type $C_{3,3}$).

Cohen's Theorem is as follows:

Theorem 8.1. Let (P,L) be a connected incidence system such that the following hold:

(F1) (P,L) is a gamma space.

(F2) If L is a line, L is not a clique.

(F3) Same as before (see section 5).

(F4) There are no minimal 5-circuits (that is any 5-circuit is derived from two edge-sharing 4-circuits).

(F5) If x and y are a polar pair and $y \in z$, then $y^{\perp} \cap z^{\perp} \neq \emptyset$.

Then (P,L) is either a nondegenerate polar space or a metasymplectic space (and hence is completely determined by the Tits theorems of section 3, as the Lie incidence system $F_{4,1}$).

A worthwhile historical note is that as a diagram geometry (that is, as a metasymplectic space) the F_4-geometries were already classified in [24], without appeal to Theorem 3.2. (In fact Tits' axiomatic treatment of polar spaces was already quite close to an M-geometry characterization.)

Cohen's proof has two parts. First since (F3) holds, Cooperstein's theory gives us a family S of symps. But because of (F5), it soon emerges that every member of S has rank 3. If p is a point and S is a symp, then $p^{\perp} \cap S$ is either empty, or is a line. As a consequence, looking at a point p, one soon discovers that the residual (L_p, Π_p, S_p) of a point is a classical near hexagon, so that the axioms (M1)-(M3) are satisfied.

9 COHEN'S CHARACTERIZATION OF THE GRASSMANN SPACES

A Grassmann space is an incidence system (P,L) such that the following four axioms hold:

(P,L) is a gamma space. (G1)

$(\mathcal{P},\mathcal{L})$ is connected but is not complete - that is some pair of non-collinear points exists. (G2)

For two points x,y with $d(x,y) = 2$, the subset $x^{\perp} \cap y^{\perp}$ is a subspace which is a generalized quadrangle. (G3)

If x is a point and L is a line such that $x^{\perp} \cap L$ is empty but $x^{\perp} \cap L^{\perp}$ is non-empty, then in fact $x^{\perp} \cap L^{\perp}$ is a line. (G4)

Noting that (G3) is a special case of the axiom (F3) encountered in section 5 and that $(\mathcal{P},\mathcal{L})$ is a gamma space, Cooperstein's theory applies. Thus a system $\mathcal{S}$ of symps exists. It is also easy to show (without using (G4)) that every line lies in a symp and so in fact $(\mathcal{P},\mathcal{L})$ is a linear incidence system. Thus $L^{\perp}$ is never a clique for any line L. If V is the singular subspace generated by three mutually collinear points not on a line, and L is a line on two of these points, then as L is not a clique, it is not V (which by (G3) must be a singular subspace). This means there is a point $d \in L^{\perp} - V^{\perp}$. Then V lies in the symp generated by d and the third point not on L among the original three points generating V. Thus every 3-clique lies in some symp and so (recalling section 5) all singular subspaces of $(\mathcal{P},\mathcal{L})$ are projective spaces.

The main characterization theorem of this section is:

Theorem 9.1. Let $(\mathcal{P},\mathcal{L})$ be a Grassmann space with no thin lines whose singular subspaces have finite rank. Then one of the following holds:

(i) $(\mathcal{P},\mathcal{L})$ is a non-degenerate polar space of rank 3.

(ii) $(\mathcal{P},\mathcal{L})$ is the Lie incidence geometry $A_{n,d}$.

(iii) For some infinite division ring K and integer $d \geq 5$ and automorphism σ of $A_{2d-1,d}(K)$ induced by a polarity of the underlying projective space of Witt index at most $d - 5$, $(\mathcal{P},\mathcal{L}) \simeq A_{2d-1,d}(K)/\langle\sigma\rangle$.

A partial characterization of finite Grassman spaces was first given by Cooperstein [11]. Later, Arjeh Cohen [8], using many of

Cooperstein's ideas, obtained a complete characterization of the Grassman spaces.

One of the main stumbling blocks in the proof is to show that each line lies in exactly two maximal subspaces. If not, there is a very special configuration which results, involving a generalized quadrangle Q and a spread R. Any two lines in R generate a grid in Q one of whose parallel classes is drawn from R. This induces a linear incidence system on R which is a projective space. But such a quadrangle Q also must appear as a point-residual in a non-degenerate rank 3 polar space S. Tits' classification of the latter is invoked to show no such Q exists.

Once we know each line lies on two maximal singular subspaces, certain affinities to Theorem 6.1 are evident, but the theorem can't be invoked, for it is not clear that, although each line lies on two maximal singular subspaces, there are really two families Σ and Π of maximal singular subspaces with the right intersection properties. Indeed they may be "fused" as one family and this is why conclusion (iii) arises. Thus the inductive machinery in Cohen's proof, utilizing two graphs $(M,\sim)$ and $(M,\tilde{\sim})$ on the set of maximal singular subspaces, must be carefully reconstructed.

10 THE COHEN-COOPERSTEIN THEOREMS

We now come to the tour de force of this survey. The spaces involved do fit under a general class described in the following definition: A parapolar space is a connected incidence system (P,L) satisfying the following three axioms:

(P,L) is a gamma space. (F1)

For any line L, $L^{\perp}$ is not a clique. (F2)

If $x,y \in P$ with $d(x,y) = 2$, then either $x^{\perp} \cap y^{\perp}$ is a single point or $x^{\perp} \cap y^{\perp}$ is a subspace isomorphic to a polar space of rank at least 2. (F3)

We have met (F1) and (F3) in section 5 outlining Cooperstein's theory, so a system S of symps is present. Also, because of (F2), we see that (P,L) is a linear incidence system and that all singular subspaces of (P,L) are projective spaces. As in section 5, we call a pair of points x,y with $d(x,y) = 2$ a special pair if $|x^{\perp} \cap y^{\perp}| = 1$ and a polar pair (called a <u>symplectic pair</u> in [9]) if $x^{\perp} \cap y^{\perp}$ is a non-degenerate polar space.

Certain specializations of the axiom (F3) will be required:

If $x,y \in P$ with $d(x,y) = 2$, then either $|x^{\perp} \cap y^{\perp}| = 1$ or $x^{\perp} \cap y^{\perp}$ is a polar subspace of rank k. $(F3)_k$

If $x,y \in P$ with $d(x,y) = 2$ then $x^{\perp} \cap y^{\perp}$ is a polar subspace of rank k. $(P3)_k$

Thus in $(F3)_k$ the rank k is specified and in $(P3)_k$ the assumption that no special pairs occur is added. We require two further axioms:

If x,y is a polar pair and L is a line on y such that $L \cap x^{\perp}$ is empty, then $x^{\perp} \cap L^{\perp}$ is either a single point or is a maximal singular subspace of $x^{\perp} \cap y^{\perp}$. (F4)

If x,y is a polar pair and L is a line on y such that $L \cap x^{\perp}$ is empty, then $x^{\perp} \cap L^{\perp}$ is either empty or is a maximal singular subspace $x^{\perp} \cap y^{\perp}$. (P4)

Axiom (F4) is equivalent to saying that for any point-symp pair $(p,S) \in P \times S$ with $p \notin S$, $p^{\perp} \cap S$ is either empty, a <u>line</u>, or a maximal singular subspace of S. (P4) says that $p^{\perp} \cap S$ is either empty, a <u>point</u>, or a maximal singular subspace of S.

Theorem 10.1 (Cohen and Cooperstein [9]) Let $k \geq 2$ and let (P,L) be a parapolar space with no thin lines, whose singular subspaces have finite rank s. Then (P,L) satisfies $(P3)_k$ and (P4) if and only if one of the following holds:

(i) $k = s$ and (P,L) is a non-degenerate polar space of rank $k + 1$ with thick lines.

(iia) $k = 2$, $s \geq 3$ and for some natural number n between 4 and $2s - 1$, and division ring K, $(P,L) \simeq A_{n,d}(K)$, $d = n - s + 1$.

(iib) $k = 2$, $s \geq 5$ and $(P,L) \simeq A_{2s-1,s}(K)/\langle\sigma\rangle$ as in conclusion (iii) of Theorem 9.1.

(iii) $k = 3$, $s \geq 4$, and for some field F, (P,L) contains families S and D of convex subspaces isomorphic to $D_{4,1}(F)$ and $D_{5,5}(F)$ respectively such that S is the system of symps of the parapolar space and if $(x,S) \in P \times S$ with $x \notin S$ and $x^{\perp} \cap S$ a maximal singular subspace of S (as provided by (P4)) then $\{x\} \cup S$ lies in a unique member of D. The incidence system of lines and planes lying on any point x is $A_{s,2}(F)$.

(iv) $k = 4$, $s = 5$ and $(P,L) \simeq E_{6,1}(F)$ for a field F.

(v) $k = 5$, $s = 6$ and $(P,L) \simeq E_{7,1}(F)$ for a field F.

It should be pointed out that Cooperstein has shown, via an important theorem in [12], that if (P,L) is a strong gamma space (recall this definition from section 4), then one may be more specific about the conclusion in (iii). In that case $(P,L) \simeq D_{n,n}$.

The next theorem represents a jump "upward" in that point residuals of its geometries are to be found among the "conclusion geometries" of Theorem 10.1.

Theorem 10.2. Let $k \geq 3$ and suppose (P,L) is again a parapolar space with no thin lines, and finite singular rank s. Then (P,L) satisfies $(F3)_k$ and (F4) if and only if there exists a field F such that one of the following holds:

(i) $k = s$ and (P,L) is a non-degenerate polar space of rank $k + 1$ with thick lines.

(ii) $k = 3$, $s = 4$, and $(P,L) \simeq D_{5,5}(F)$ or $E_{6,4}(F)$.

(iii) $k = 4$, $s = 5,6$ and $(P,L) \simeq E_{6,1}(F)$ or $E_{7,7}(F)$.

(iv) $k = 6$, $s = 7$ and $(P,L) \simeq E_{8,1}(F)$.

It should be apparent now that all spherical buildings have been reached, so there is a characterization of each solely in terms of points and lines. The proofs of these theorems, of course, are involved, but basically they aim at constructing various species of subspaces, obtaining along with them certain information about their intersections - enough (where possible) to invoke Tits' fundamental theorems on geometries of type M (section 2).

REFERENCES

[1] Aschbacher, M. Presheaves on Tits geometries. Preprint.
[2] Buekenhout, F. The basic diagram of a geometry. Springer Lecture Notes in Math. 893, Berlin, 1981.
[3] Buekenhout, F. Cooperstein's theory. To appear in Simon Stevin.
[4] Buekenhout, F. and Shult E. E. On the foundations of polar geometry. Geometriae Dedicata, 3 (1974), 155-170.
[5] Buekenhout, F. and Sprague, A. Personal communication.
[6] Cameron, P. Dual polar spaces. Geometriae Dedicata 12 (1982), 75-85.
[7] Cohen, A. An axiom system for metasymplectic spaces. Geometriae Dedicata 12 (1982), 417-433.
[8] Cohen, A. On a theorem of Cooperstein. To appear in European J. Comb.
[9] Cohen, A. and Cooperstein, B. A characterization of some geometries of exceptional Lie type. Preprint.
[10] Cooperstein, B. Some geometries associated with parabolic representations of groups of Lie type. Can. J. Math. 28 (1976), 1021-1031.
[11] Cooperstein, B. A characterization of some Lie incidence structures. Geometriae Dedicata 6 (1977), 205-258.
[12] Cooperstein, B. A characterization of a geometry related to $\Omega^+_{2m}(K)$. To appear in J. Comb. Theory.
[13] Freudenthal, H. Beziehungen der E_7 and E_8 zur Oktavenebene, I-XI. Proc. Kon. Ned. Akad. Wet. A57 (1954), 218-230, 363-368; A58 (1955), 151-157, 227-285; A62 (1959), 165-201, 447-474; A66 (1963), 457-487 (= Indag. Math., 16, 17, 21, 25).
[14] Higman, D. G. Invariant relations, coherent configurations and generalized polygons, pp 27-43 in "Combinatorics, part 3: Combinatorial group theory", Math Centre Tract 57, Amsterdam, 1974.
[15] Johnson, P. and Shult, E. Gamma spaces which are locally polar spaces. Submitted to Geometriae Dedicata.
[16] Ronan, M. Coverings and automorphisms of chamber systems. European J. Comb. 1 (1980), 259-269.
[17] Ronan, M. Locally truncated buildings and M_{24}. To appear in Math. Zeit.
[18] Ronan, M. Extending locally truncated buildings and chamber systems. Preprint.
[19] Shad, S. and Shult, E. The near n-gon geometries. Preprint.
[20] Shult, E. Groups, polar spaces and related structures, in Proceedings of the Advanced Study Institute on Combinatorics, Breuklen, 1975, eds. M. Hall Jr. and J. H. van Lint, Mathematical Centre Tracts no. 55, Amsterdam, 1974, 130-161.
[21] Shult, E. and Yanushka, A. Near n-gons and line systems. Geometriae Dedicata 9 (1980), 1-72.
[22] Sprague, A. Dual linear, linear diagrams, in Finite Geometries, Lecture Notes in Pure and Applied Math. 82, ed. N. Johnson, M. Kallaher and C.T. Long, Marcel Dekker, 1982.
[23] Tallini, G. On a characterization of the Grassman manifold representing the lines of a projective space, pp 354-358 in Finite Geometries and Designs, proceedings of the second Isle of Thorns Conference, 1980, London Math. Soc. Lecture Note 49, Cambridge University Press 1981.

[24] Tits, J. Buildings of spherical type and finite BN pairs. Springer Lecture Notes in Math. 386, Berlin, 1974.

[25] Tits, J. A local approach to buildings, in The Geometric Vein, ed. C. Davis, B. Grünbaum and F.A. Sherk, Springer-Verlag, 1981.

GL(n, C) FOR COMBINATORIALISTS

R. P. Stanley
Department of Mathematics
Massachusetts Institute of Technology
Cambridge, Massachusetts 02139 U.S.A.

1. INTRODUCTION

Let $G_n = GL(n, \mathbb{C}) = GL(V_n)$ denote the group of all invertible linear transformations $A: V_n \to V_n$, where V_n is an n-dimensional complex vector space. Once we choose a basis for V_n we can regard G_n as the group of nonsingular $n \times n$ complex matrices. A (linear) representation of G_n of dimension m is a homomorphism $\phi: G_n \to G_m$. We call ϕ a polynomial (respectively, rational) representation if (after choosing bases) the matrix entries of $\phi(A)$ are fixed polynomials (respectively, rational functions) in the matrix entries of A. For instance, the map

$$\begin{bmatrix} a & b \\ c & d \end{bmatrix} \xrightarrow{\phi} \begin{bmatrix} a^2 & 2ab & b^2 \\ ac & ad+bc & bd \\ c^2 & 2cd & d^2 \end{bmatrix} \qquad (1)$$

is a polynomial representation of G_2 of dimension three, while

$$\begin{bmatrix} a & b \\ c & d \end{bmatrix} \xrightarrow{\rho} (ad-bc)^{-1} \qquad (2)$$

is a rational representation of dimension one. Henceforth, all representations in this paper are assumed to be rational.

The theory of such representations has close connections with combinatorics, and our object here is to present an overview of this subject from the combinatorial viewpoint. We first will state without proof the basic results (which may be gleaned from such sources as Hamermesh (1962), pp. 377-391; James & Kerber (1981), Ch. 8; Littlewood (1950), Ch. X; Macdonald (1979), pp. 74-84; and Weyl (1946), Ch. IV), and then proceed to the combinatorial ramifications.

The first result we need is that the (rational) representations of G_n are completely reducible (i.e., G_n is a reductive group). This means in effect that every representation $\phi: G_n \to G_m$ can be decomposed into irreducibles, i.e., if $G_m = GL(V)$, so that we may regard G_n as acting on V, then $V = V_1 \oplus \cdots \oplus V_r$ where each V_i is nonzero and invariant under G_n, and no V_i has a proper G_n-invariant subspace. Although the V_i's need not be unique, the multiset $\{\phi_1,\dots,\phi_n\}$ of irreducible representations $\phi_i : G_n \to GL(V_i)$ is unique up to equivalence. Thus to determine ϕ up to equivalence, it suffices to describe the multiplicity of each irreducible representation of G_n in ϕ.

Suppose $A \in G_n$ has eigenvalues $\theta_1,\dots,\theta_n$. Then there exists a multiset M_ϕ of m Laurent monomials $u(x) = x_1^{a_1}\dots x_n^{a_n}$, $a_i \in \mathbb{Z}$, independent of A, such that the eigenvalues of $\phi(A)$ are given by the multiset $\{u(\theta_1,\dots,\theta_n) \mid u \in M_\phi\}$. For the representations ϕ and ρ of (1) and (2), the reader can check that $M_\phi = \{x_1^2, x_1x_2, x_2^2\}$ and $M_\rho = \{x_1^{-1}x_2^{-1}\}$. The Laurent polynomial $f_\phi = \sum_{u \in M_\phi} u$ (which is a symmetric function of $x_1,\dots,x_n$) is called the <u>character</u> of ϕ; clearly

$$f_\phi(\theta_1,\dots,\theta_n) = \mathrm{tr}\ \phi(A),$$

where tr denotes trace. The character f_ϕ uniquely determines ϕ up to equivalence. In other words, f_ϕ can be written uniquely as a nonnegative integral combination of irreducible characters. We now wish to describe the irreducible characters of G_n. First we reduce this problem to polynomial representations.

1.1 <u>Theorem</u>. Any rational representation $\phi : GL_n \to GL_m$ has the form $\phi(A) = (\det A)^{-r}\phi'(A)$ for some $r \in \mathbb{Z}$ and some polynomial representation ϕ'. ϕ is irreducible if and only if ϕ' is irreducible, and $f_\phi(x_1,\dots,x_n) = (x_1\dots x_n)^{-r}\ f_{\phi'}(x_1,\dots,x_n)$.

Now let $\lambda = (\lambda_1,\dots,\lambda_n)$ be any partition into $\leq$ n parts, i.e., $\lambda_1 \geq \dots \geq \lambda_n \geq 0$, $\lambda_i \in \mathbb{Z}$. The number of (positive) parts $\lambda_i > 0$ of λ is called the <u>length</u> of λ, denoted $\ell(\lambda)$. We also write $|\lambda| = \lambda_1 + \dots + \lambda_n$. Following Macdonald (1979), let $s_\lambda(x_1,\dots,x_n)$ denote the <u>Schur function</u> corresponding to λ in the variables $x_1,\dots,x_n$. It has the following combinatorial definition. Write down a left-justified array whose entries are the numbers 1,2,...,n (with any multiplicities), with λ_i entries in row i, such that the columns are strictly decreasing and rows weakly decreasing. With such an array T (called a <u>tableau</u> of <u>shape</u> λ and largest part $\leq$ n) associate the monomial $m(T) = x_1^{a_1}\dots x_n^{a_n}$, where a_i i's appear in T. Then $s_\lambda(x) = s_\lambda(x_1,\dots,x_n)$ is defined to be $\sum_T m(T)$, summed over all tableaux T of shape λ and largest part $\leq n$. Though not obvious from the definition, $s_\lambda(x)$ is a symmetric function of $x_1,\ldots,x_n$.

Example. Take $\lambda = (2,1)$, $n = 3$. The appropriate tableaux are

$$\begin{matrix} 21 & 22 & 31 & 33 & 32 & 33 & 32 & 31 \\ 1 & 1 & 1 & 1 & 2 & 2 & 1 & 2 \end{matrix} .$$

Hence $s_{21}(x_1,x_2,x_3) = x_1^2x_2+x_1x_2^2 + x_1^2x_3 + x_1x_3^2 + x_2^2x_3 + x_2x_3^2 + 2x_1x_2x_3$.

The main result on the polynomial characters of G_n is the following.

1.2 Theorem. Let $\lambda = (\lambda_1,\ldots,\lambda_n)$ be a partition, $\ell(\lambda)\leq n$. Then the Schur function $s_\lambda(x_1,\ldots,x_n)$ is an irreducible polynomial character of GL_n, different λ's yield different characters, and every irreducible polynomial character has this form.

We will denote by ρ_λ the representation of G_n whose character is s_λ. In other words, $s_\lambda = f_{\rho_\lambda}$.

The Schur functions $s_\lambda(x_1,\ldots,x_n)$, $\ell(\lambda)\leq n$, form a $\mathbb{Z}$-basis for the additive group of all symmetric polynomials in $x_1,\ldots,x_n$ with integer coefficients (Macdonald 1979, p.24). Thus every such polynomial f is a virtual character (= difference of two characters) of G_n, and expanding f in terms of Schur functions is equivalent to finding the multiplicity of each irreducible character of G_n in f.

Remarks on some other groups. The representations of the groups U(n), SL(n,ℂ), and SU(n) can be obtained easily from those of G_n = GL(n,ℂ). Since G_n is a reductive algebraic group with maximal compact subgroup U(n) it follows from general principles that the rational representations of G_n and U(n) coincide. More precisely, distinct irreducible representations of G_n restrict to distinct irreducibles of U(n) and every irreducible representation of U(n) arises in this way.

Regarding SL(n,ℂ), suppose $\phi: G_n \to G_m$ has character $s_\lambda(x)$. By our definition of s_λ,

$$s_\lambda(x) = (x_1\ldots x_n)^{\lambda_n} s_{\lambda^*}(x), \tag{3}$$

where $\lambda^* = (\lambda_1-\lambda_n, \lambda_2-\lambda_n,\ldots,\lambda_{n-1}-\lambda_n, 0)$. If $\phi^*: G_n \to G_m$ has character ϕ^*, then the right-hand side of (3) is just the character of $(\det)^{\lambda_n}\phi^*$. Hence $\phi = (\det)^{\lambda_n}\phi^*$, so that ϕ and ϕ^* restrict to the same representation of SL(n,ℂ). But except for this, the irreducible representations of GL(n,ℂ) and SL(n,ℂ) coincide. More precisely:

1.3 <u>Theorem.</u> Let $\lambda = (\lambda_1,\ldots,\lambda_{n-1})$ be a partition into $\leq n-1$ parts. Then the Schur function $s_\lambda(x_1,\ldots,x_n)$ is an irreducible polynomial character of $SL(n,\mathbb{C})$, different λ's yield different characters, and every irreducible polynomial character has this form.

We will call any Laurent polynomial $f(x_1,\ldots,x_n)$ a <u>character</u> of the representation $\rho: SL(n,\mathbb{C}) \to G_m$ if $\operatorname{tr} \rho(A) = f(\theta_1,\ldots,\theta_n)$ for all $A \in SL(n,\mathbb{C})$ with eigenvalues $\theta_1,\ldots,\theta_n$. Since $\theta_1\cdots\theta_n = 1$, the character is not unique (as it was for G_n). The character f_ρ of ρ is, however, a uniquely defined element of the quotient ring $\Lambda(x_1,\ldots,x_n)/(x_1\cdots x_n - 1)$, where $\Lambda(x_1,\ldots,x_n)$ denotes the ring of symmetric polynomials with integer coefficients in the variables $x_1,\ldots,x_n$. Thus frequently we will carry out our computations with characters of $SL(n,\mathbb{C})$ in this quotient ring.

Finally, $SU(n)$ bears the same relation to $SL(n,\mathbb{C})$ as $U(n)$ does to $GL(n,\mathbb{C})$.

2. <u>SOME EXAMPLES</u>

Let us consider some "naturally occurring" representations of G_n and try to compute their characters. First we have the <u>defining representation</u> $\phi: G_n \to G_n$ given by $\phi(A) = A$. If A has eigenvalues $\theta_1,\ldots,\theta_n$ then $\phi(A)$ also has these eigenvalues. Hence $\operatorname{tr} \phi(A) = \theta_1+\cdots+\theta_n$ and $f_\phi(x) = x_1+\cdots+x_n$. Since for each $1 \leq i \leq n$ there is exactly one way to put i into the shape $\lambda = (1)$ to form a column-strict plane partition, we have $s_1(x) = x_1+\cdots+x_n$. Hence $f_\phi = s_1$ and $\phi = \phi_1$.

Suppose $\phi: G_n \to G_m = GL(V_m)$ is any representation. Choose a basis $z_1,\ldots,z_m$ for the vector space V_m. Let $S^k(V_m)$ denote the vector space of all homogeneous polynomials of degree k in the variables $z_1,\ldots,z_m$. Thus $\dim S^k(V_m) = \binom{m+k-1}{k}$, and $S^k(V_m)$ is the <u>k-th symmetric power</u> of V_m. Any $B \in G_m$ acts on $S^k(V_m)$ by the rule $B \cdot g(z_1,\ldots,z_m) = g(Bz_1,\ldots,Bz_m)$ so we have a representation of G_m on $S^k(V_m)$, i.e., a homomorphism $G_m \to GL(S^k(V_m)) \cong G_{\binom{m+k-1}{k}}$. Hence G_n acts on $S^k(V_m)$ by composition, i.e., if $A \in G_n$ and $g \in S^k(V_m)$ then $A \cdot g = \phi(A) \cdot g$. The resulting representation is denoted $S^k\phi: G_n \to GL(S^k(V_m))$. It is an important and difficult problem (which comes close to subsuming all of classical invariant theory) to decompose $S^k\phi$ into irreducibles.

The problem of decomposing $S^k\phi$ (up to equivalence) may be stated in combinatorial terms as follows. Let $A = \operatorname{diag}(\theta_1,\ldots,\theta_m) \in G_m$, with respect to the basis $z_1,\ldots,z_m$ of V_m. Write S^kA for the action of A on

S^kV_m, i.e., $S^kA = (S^k\phi_1)(A)$, where $\phi_1:G_m\to G_m$ is the defining representation. A monomial $z_1^{a_1}\cdots z_m^{a_m}\in S^kV_m$ is an eigenvector for S^kA with eigenvalue $\theta_1^{a_1}\cdots\theta_m^{a_m}$. Since the monomials $z_1^{a_1}\cdots z_m^{a_m}$ of degree k form a basis for S^kV_m, we have accounted for all the eigenvalues of S^kA. Hence

$$\begin{aligned}\operatorname{tr} S^kA &= \sum_{a_1+\cdots+a_m=k} \theta_1^{a_1}\cdots\theta_m^{a_m}\\ &= \text{coefficient of } q^k \text{ in } \prod_{i=1}^m (1-\theta_i q)^{-1}. \qquad (4)\end{aligned}$$

Let $\mathcal{M}_\phi$ be the multiset of monomials for $\phi: G_n\to G_m$ defined above, so $f_\phi(x_1,\ldots,x_n) = \sum_{u\in\mathcal{M}_\phi} u$. It follows from (4) that

$$\sum_{k\geq 0} f_{S^k\phi}(x)q^k = \prod_{u\in\mathcal{M}_\phi} (1-uq)^{-1}. \qquad (5)$$

Thus the problem of decomposing $S^k\phi$ (up to equivalence) is equivalent to the combinatorial problem of expanding the right-hand side of (5) in terms of Schur functions. This is a special case of the notion of plethysm of Schur functions; see Macdonald (1979), p.82.

For now we will be content with decomposing $S^k\phi_1$ (where $\phi_1:G_n\to G_n$ is the defining representation). By (5) we have

$$\begin{aligned}\sum_{k\geq 0} f_{S^k\phi_1}(x)q^k &= \prod_{i=1}^n (1-x_iq)^{-1}\\ &= \sum_{k\geq 0} h_k(x)q^k,\end{aligned}$$

where $h_k(x)$ is the sum of all monomials in $x_1,\ldots,x_n$ of degree k (called the complete (homogeneous) symmetric functions). For any integers $b_1,\ldots,b_n\geq 0$ satisfying $\Sigma b_i=k$ there is a unique tableau of shape (k) = (k,0,0,...) with b_i i's. Hence $f_{S^k\phi_1}(x) = h_k(x) = s_k(x)$, i.e., $S^k\phi_1$ is irreducible with character s_k. Since $f_{\phi_1} = s_1$ we can write $s_k = S^ks_1$.

In an exactly analogous way, given $\phi:G_n\to G_m = GL(V_m)$ we can compute the character of $\Lambda^k\phi$: $G_n\to GL(\Lambda^kV_m)$, where Λ^k denotes the k-th exterior power, $0\leq k\leq m$. Keeping the same notation as before, a wedge product $z_{i_1}\wedge\cdots\wedge z_{i_k}$, $1\leq i_1<\cdots<i_k\leq m$, is an eigenvector for Λ^kA with eigenvalue $\theta_{i_1}\cdots\theta_{i_k}$. Hence

$$\sum_{k=0}^{m} f_{\Lambda^k\phi}(x)q^k = \prod_{u\in M_\phi} (1+uq). \qquad (6)$$

Letting $\phi = \phi_1$ we obtain

$$\sum_{k=0}^{m} f_{\Lambda^k\phi_1}(x)q^k = \prod_{i=1}^{n} (1+x_i q)$$
$$= \sum_{k=0}^{m} e_k(x)q^k,$$

where $e_k(x)$ denotes the k-th <u>elementary symmetric function</u> in $x_1,\ldots,x_n$. For any integers $1\le c_1<\ldots<c_k\le n$ there is a unique tableau of shape $(1^k) = (1,1,\ldots,1)$ (k ones) with parts $c_1,\ldots,c_k$. Hence $f_{\Lambda^k\phi}(x)=e_k(x)=s_{1^k}(x)$, i.e. $\Lambda^k\phi_1$ is irreducible with character s_{1^k}, and we can write $s_{1^k} = \Lambda^k s_1$.

Let us compute one additional example of this nature, which will be of use in Section 4. Let M_n denote the n^2-dimensional vector space of all n×n matrices. Then $A\in G_n$ acts on M_n by the rule $B\to A^{-1}BA$, where $B\in M_n$. This representation of G_n is called the <u>adjoint representation</u>, denoted ad. We now compute its character. Let E_{ij} be the elementary matrix with a one in position (i,j) and zeros elsewhere. Choose $A=\mathrm{diag}(\theta_1,\ldots,\theta_n)$. Then $A^{-1}E_{ij}A = \theta_i^{-1}\theta_j E_{ij}$. Thus

$$\mathrm{tr}(\mathrm{ad}\ A) = \sum_{i,j} \theta_i^{-1}\theta_j$$
$$= (\theta_1\cdots\theta_n)^{-1} \sum_{i,j} (\theta_1\cdots\theta_n)\theta_i^{-1}\theta_j. \qquad (7)$$

Consider the partition $(2,1,\ldots,1)\vdash n$. To form a tableau of this shape with largest part $\le n$, choose any (n-1)-element subset S of $\{1,\ldots,n\}$ and insert it (uniquely) into the first column. The additional entry can be any element of $\{1,\ldots,n\}$, with the sole exception that it cannot equal n when $S = \{1,\ldots,n-1\}$. It follows that

$$s_{21^{n-2}}(x_1,\ldots,x_n) = \sum_{i,j} (x_1\ldots x_n)x_i^{-1}x_j - (x_1\ldots x_n).$$

Comparison with (7) yields

$$f_{ad} = (\det)^{-1}s_{21^{n-2}} + s_{\emptyset} \qquad (\emptyset = \text{null set}).$$

Since $s_{\emptyset}$ is the character of the trivial representation, this means that M_n has a (unique) one-dimensional subspace W fixed pointwise by G_n. Of course W is just the set of scalar matrices. The complementary invariant subspace M_n^0 to W (the one which affords the character $(\det)^{-1}s_{21^{n-2}}$) consists of the matrices of trace 0. If we restrict the action of G_n on M_n^0 to SL(n,ℂ), then we obtain the adjoint representation of SL(n,ℂ), with character

$$f_{ad} = s_{21^{n-2}}(x_1,\ldots,x_n) = \sum_{i\neq j} (x_1\ldots x_n)x_i^{-1}x_j + (n-1)(x_1\ldots x_n).$$

Since characters of SL(n,ℂ) are defined modulo the relation $x_1\ldots x_n = 1$, we could also write

$$f_{ad} = \sum_{i\neq j} x_i^{-1}x_j + n-1. \tag{8}$$

Though this is not in the "canonical form" given by Theorem 1.3, it is an equally valid expression for f_{ad} (and one which is more natural from the Lie-algebraic viewpoint).

As an exercise, the reader may wish to compute the characters of the actions of G_n on M_n given by BA, $A^{-1}B$, BA^*, A^tB, A^tBA, $A^{-1}BA^*$, and A^tBA^*, where t denotes transpose and $A^* = (A^t)^{-1}$.

Virtually any identity involving symmetric functions can be interpreted in terms of representation theory. We give one such example here.

An elegant combinatorial proof (Knuth 1970, Stanley 1971) can be given of the identity

$$\prod_{i=1}^{n} (1-x_i)^{-1} \prod_{\substack{i,j=1 \\ i<j}}^{n} (1-x_ix_j)^{-1} = \sum_{\substack{\lambda \\ \ell(\lambda)\leq n}} s_\lambda(x_1,\ldots,x_n). \tag{9}$$

Now $s_1(x) = \Sigma x_i$ and $s_{11}(x) = \Lambda^2 s_1(x) = \sum_{i<j} x_ix_j$. Thus the left-hand side is the character of the representation $S(\rho_1+\rho_{11})$, i.e., the natural action of $G_n = GL(V_n)$ on the symmetric algebra $S(V_n \oplus \Lambda^2 V_n)$. Thus by (9) we see that in the representation $S(\rho_1+\rho_{11})$, every irreducible polynomial representation of G_n occurs exactly once. A refinement of (9)

(Macdonald 1979, p.46, Ex.7) asserts that

$$\prod_i (1-tx_i)^{-1} \prod_{i<j} (1-x_ix_j)^{-1} = \sum_\lambda t^{c(\lambda)} s_\lambda(x),$$

where $c(\lambda)$ is the number of columns of odd length in λ. From this it is easy to obtain the decomposition of each $S^k(\rho_1+\rho_{11})$, viz., ρ_λ appears in $S^k(\rho_1+\rho_{11})$ (with multiplicity one) if and only if $k = \frac{1}{2}(|\lambda|+c(\lambda))$.

3. UNIMODALITY

Consider the group $SL(2,\mathbb{C})$. By Theorem 1.3 the irreducible characters are just the Schur functions

$$s_m(x,y) = x^m+x^{m-1}y+\cdots+y^m .$$

(Thus the irreducible representations are just $S^m(\rho_1)$.) It is more usual to write this character as

$$s_m(x,x^{-1}) = x^{-m} + x^{-m+2}+\cdots+x^m,$$

which of course is the same as before modulo the relation $xy = 1$. Now suppose ρ is any representation and that ρ_m appears in ρ with multiplicity a_m. Then for sufficiently large k, ρ has the character

$$f_\rho(x,x^{-1}) = \sum_{m=0}^{k} a_m(x^{-m}+x^{-m+2}+\cdots+x^m)$$

$$= \sum_{j=-k}^{k} b_jx^j ,$$

where $b_j = a_j+a_{j+2}+a_{j+4}+\cdots$ for $j\geq 0$, and $b_j = b_{-j}$. It follows that $b_0\geq b_2\geq b_4\geq\cdots$ and $b_1\geq b_3\geq\cdots$. We say that a sequence $c_0,c_1,\ldots,c_r$ is unimodal if $c_0\leq c_1\leq\cdots\leq c_s$ and $c_s\geq c_{s+1}\geq\cdots\geq c_r$ for some s, and is symmetric if $c_i = c_{r-i}$. Thus we have shown:

3.1 Theorem. For any representation ρ of $SL(2,\mathbb{C})$ with character $f_\rho(x,x^{-1}) = \sum_{j=-k}^{k} b_jx^j$, the two sequences $b_{-k},b_{-k+2},\ldots,b_k$ and $b_{-k+1},b_{-k+3},\ldots,b_{k-1}$ are symmetric and unimodal.

Theorem 3.1 can be used as a tool in showing that certain sequences of combinatorial interest are unimodal. For a general discussion of this topic, see Almkvist (1982). Here we present the prototypical case, viz., the action of $SL(V_2)$ on $S^k(S^mV_2)$ or equivalently, the representation

$S^k\rho_m$.

Since $s_m(x,x^{-1}) = x^{-m}+x^{-m+2}+\cdots+x^m$, the character of $\rho = S^k\rho_m$ is given by

$$f_\rho(x,x^{-1}) = \Sigma\ (x^{-m})^{r_0}(x^{-m+2})^{r_1}\cdots(x^m)^{r_m}$$
$$= x^{-mk}\ \Sigma(x^2)^{r_1+2r_2+\cdots+mr_m}$$

where the sums range over all nonnegative integral sequences $(r_0,r_1,\ldots,r_m)$ satisfying $r_0+r_1+\cdots+r_m = k$. Identify $(r_0,r_1,\ldots,r_m)$ with the partition λ with r_i parts equal to i. Then the sum ranges over all partitions λ whose Ferrers diagram (e.g., Andrews 1976, p.6) fits in a k×m rectangle, and $r_1+2r_2+\cdots+mr_m = |\lambda|$. Hence the coefficient of x^{-mk+2j} in $f_\rho(x,x^{-1})$ is equal to the number $p(j,m,k)$ of partitions of j fitting in a k×m rectangle (i.e., with $\leq$k parts and largest part $\leq$m). It follows from Theorem 3.1 that for fixed m and k, the sequence

$$p(0,m,k),\ p(1,m,k),\ldots,\ p(mk,m,k)$$

is symmetric and unimodal. Although it is possible to prove this result without mentioning SL(2,ℂ) (e.g., Stanley 1982, Cor. 9.6), no simple combinatorial proof is known.

Let us also mention that the polynomial $\sum_j p(j,m,k)q^j$ is the <u>q-binomial coefficient</u> $\left[{m+k \atop k}\right]$, defined by

$$\left[{m+k \atop k}\right] = \frac{[m+k]!}{[m]!\,[k]!}$$

where $[i]! = (1-q)(1-q^2)\cdots(1-q^i)$. (See Andrews 1976, Thm. 3.1) Thus we have shown that the coefficients of $\left[{m+k \atop k}\right]$ are symmetric and unimodal.

4. A LITTLE INVARIANT THEORY

If a group G acts on a ring R, then the fixed ring

$$R^G = \{f\in R \mid Af = f \text{ for all } A\in G\}$$

is called the <u>ring of invariants</u> of G. Suppose we're given a decomposition $R = R_0 \oplus R_1 \oplus \cdots$, where each R_i is a finite-dimensional vector space over a field K and $\oplus$ denotes vector space direct sum. Suppose G acts so that $GR_i = R_i$. Then $R^G = R_0^G \oplus R_1^G \oplus \cdots$, where $R_i^G = R_i \cap R^G$. We then call the power series

$$F(R^G,q) = \sum_{i\geq 0}(\dim R_i^G)q^i$$

the Molien series of R^G (or of G acting on R).

Consider the special case $G = SL(n,\mathbb{C})$ and $R = S(V) = \mathbb{C} \oplus S^1(V) \oplus S^2(V) \oplus \cdots$, where V affords the adjoint representation of $SL(n,\mathbb{C})$ (i.e., V is the space of n×n complex matrices of trace 0, and G acts on V by conjugation). It is easy to see that R^G is generated by n-1 algebraically independent elements $\theta_1,\ldots,\theta_{n-1}$ of degrees 2,3,...,n. Namely, for $B\in V$ take $\theta_i(B)$ to be the coefficient of λ^{n-i-1} in the characteristic polynomial of B. It follows that

$$F(S(V)^G,q) = 1/(1-q^2)\cdots(1-q^n). \tag{10}$$

We will give a combinatorial derivation of (10).

Since we are working with $SL(n,\mathbb{C})$, we deal with the variables $x = (x_1,\ldots,x_n)$ and all our computations are performed modulo the relation $x_1\cdots x_n = 1$. Let $\lambda = (\lambda_1,\ldots,\lambda_{n-1})$ be a partition, and define $\tilde{\lambda} = (\lambda_1, \lambda_1-\lambda_{n-1}, \lambda_1-\lambda_{n-2},\ldots,\lambda_1-\lambda_2)$. We claim

$$s_\lambda(1/x_1,\ldots,1/x_n) = s_{\tilde{\lambda}}(x_1,\ldots,x_n) \tag{11}$$

(always modulo $x_1\cdots x_n=1$). While it is easy to give a representation-theoretic proof of (11), a combinatorial approach is instructive. Namely, given a tableau T of shape λ and largest part $\leq n$, define a new tableau $\tilde{T}$ of shape $\tilde{\lambda}$ by the following condition: if $a_1,\ldots,a_k$ are the elements of the i-th column of T, then the elements $b_1,\ldots,b_{n-k}$ of column λ_1-i+1 of $\tilde{T}$ consist of the elements of the complementary set $\{1,\ldots,n\}-\{a_1,\ldots,a_k\}$. For instance, if n=4, then we have

$$T = \begin{matrix} 4&4&3&3&3&2&1&1 \\ 3&2&2&1 \\ 1&1 \end{matrix} \qquad \tilde{T} = \begin{matrix} 4&4&4&4&4&4&3&2 \\ 3&3&3&2&2&1 \\ 2&2&1&1 \end{matrix}.$$

This sets up a one-to-one correspondence between the terms of $s_{\tilde{\lambda}}(x_1,\ldots,x_n)$ and $(x_1 \cdots x_n)^{\lambda_1} s_\lambda(1/x_1,\ldots,1/x_n)$, so (11) follows.

Next consider the product

$$(1-q)^{-n+1} \prod_{i\neq j} (1-qx_ix_j^{-1})^{-1} = \sum_{\lambda=(\lambda_1,\ldots,\lambda_{n-1})} P_\lambda(q)s_\lambda(x), \tag{12}$$

where $P_\lambda(q)$ is a formal power series in q. It follows from (5) and (8) that $F_G(q) = P_\emptyset(q)$. In order to expand the left-hand side of (12), we begin with the identity

$$\prod_{i,j=1}^{n} (1-x_i y_j)^{-1} = \sum_{\lambda=(\lambda_1,\ldots,\lambda_n)} s_\lambda(x) s_\lambda(y), \qquad (13)$$

which can be given an elegant combinatorial proof (Knuth 1970, p.726; Stanley 1971, Cor.7.2) similar to that of (9). Make the substitution $y_j \to q x_j^{-1}$ in (13). We obtain

$$(1-q)^{-n} \prod_{i\neq j} (1-qx_i x_j^{-1})^{-1} = \sum_\lambda q^{|\lambda|} s_\lambda(x) s_\lambda(1/x)$$
$$= \sum_\lambda q^{|\lambda|} s_\lambda(x) s_{\tilde{\lambda}}(x) \quad \text{(by (11))}. \qquad (14)$$

(This is similar to Macdonald (1979), ex.5, p.37.) We now appeal to the Littlewood-Richardson rule (Macdonald 1979, I.9) for multiplying Schur functions. It is easy to deduce from this rule (see Stanley 1977, Thm.3.4) that for any partitions μ and ν into $\leqslant n$ parts, when we expand $s_\mu(x)s_\nu(x)$ in terms of $s_\rho(x)$ for $\ell(\rho) \leqslant n-1$ (working modulo $x_1 \cdots x_n = 1$ as always), the coefficient $s_\emptyset(x)$ is $\delta_{\mu\tilde{\nu}}$. In particular, the coefficient of $s_\emptyset(x)$ in $s_\lambda(x)s_{\tilde{\lambda}}(x)$ is one. Thus from (14) we obtain

$$(1-q)^{-1} P_\emptyset(q) = \sum_{\lambda=(\lambda_1,\ldots,\lambda_n)} q^{|\lambda|}$$
$$= 1/(1-q)(1-q^2)\ldots(1-q^n) \ ,$$

so (10) follows.

An analogous argument applies to the case $G = GL(n,\mathbb{C})$ and $R = \Lambda(V) = \mathbb{C} \oplus \Lambda(V) \oplus \cdots \oplus \Lambda^{n^2-1}(V)$, where once again V affords the adjoint representation. Instead of (13) we use

$$\prod_{i,j=1}^{n} (1+x_i y_j) = \sum_{\lambda=(\lambda_1,\ldots,\lambda_n)} s_\lambda(x) s_{\lambda'}(y) \ , \qquad (15)$$

where λ' denotes the conjugate partition to λ (see Knuth 1970, p.726; Stanley 1971, Cor.9.2). Thus we obtain

$$(1+q)^n \prod_{i\neq j} (1+qx_i x_j^{-1}) = \sum_\lambda q^{|\lambda|} s_\lambda(x) s_{\tilde{\lambda}'}(x) \ .$$

The coefficient of $s_\emptyset(x)$ in $s_\lambda(x)s_{\tilde{\lambda}'}(x)$ is one if $\lambda = \lambda'$ and otherwise zero. Hence in this case by (6) and (8) we have

$$(1+q)F(\Lambda(V)^G,q) = \sum_{\substack{\lambda=\lambda' \\ \lambda=(\lambda_1,\ldots,\lambda_n)}} q^{|\lambda|} . \tag{16}$$

It is well-known (and easy to prove combinatorially - see Hardy & Wright (1960), pp.278-9) that the number of self-conjugate partitions of m with $\leq n$ parts is equal to the number of partitions of m into distinct odd parts $\leq 2n-1$. Thus the right-hand side of (16) becomes $(1+q)(1+q^3)\ldots(1+q^{2n-1})$, so

$$F(\Lambda(V)^G,q) = (1+q^3)(1+q^5)\cdots(1+q^{2n-1}). \tag{17}$$

This well-known result is usually proved by much more algebraic means (e.g., Weyl 1946, p.233; Kostant 1958).

The "q-Dyson conjecture" of Andrews (1975), p.216 (see also Macdonald (1981,1982)), in the case $a_1=\cdots=a_n=k$, is equivalent to finding the coefficient of $s_\emptyset(x)$ in

$$\prod_{m=1}^{k} \prod_{i,j=1}^{n} (1-q^m x_i x_j^{-1}).$$

Perhaps the combinatorial techniques illustrated here will be of value in resolving the conjecture.

<u>Late note</u>: Our proof of formula (17) is essentially that of D. E. Littlewood (1953), On the Poincaré polynomials of the classical groups, J. London Math. Soc., <u>28</u>, 494-500.

REFERENCES

Almkvist, G. (1982), Representations of SL(2,C) and unimodal polynomials. Department of Mathematics, University of Lund.

Andrews, G.E. (1975). Problems and prospects for basic hypergeometric functions. In Theory and Application of Special Functions, pp. 191-224. New York: Academic Press.

Andrews, G.E. (1976). The Theory of Partitions. Reading, Mass.: Addison-Wesley.

Hamermesh, M. (1962). Group Theory and Its Application to Physical Problems. Reading, Mass.: Addison-Wesley.

Hardy, G.H. & Wright, E.M. (1960). An Introduction to the Theory of Numbers, fourth ed. Oxford: Oxford University Press.

James, G. & Kerber, A. (1981). The Representation Theory of the Symmetric Group. Reading, Mass.: Addison-Wesley.

Knuth, D.E. (1970). Permutations, matrices, and generalized Young tableaux. Pacific J. Math., 34, 709-727.

Kostant, B. (1958). A theorem of Frobenius, a theorem of Amitsur-Levitski and cohomology theory. J. Math. Mech., 7, 237-64.

Littlewood, D.E. (1950). The Theory of Group Characters, second ed. Oxford: Oxford University Press.

Macdonald, I.G. (1979). Symmetric Functions and Hall Polynomials. Oxford: Oxford University Press.

Macdonald, I.G. (1981). Some conjectures for root systems and finite Coxeter groups. In Sem. d'Alg. Paul Dubriel et Marie-Paule Malliavin, Lecture Notes in Math., no. 867, pp. 90-97. Berlin: Springer.

Macdonald, I.G. (1982). Some conjectures for root systems and finite Coxeter groups. SIAM J. Math. Anal., 13, (1982), 988-1007. (Expanded version of above reference.)

Stanley, R.P. (1971). Theory and application of plane partitions, Parts 1 and 2. Studies in Applied Math., 50, 167-188, 259-79.

Stanley, R.P. (1977). Some combinatorial aspects of the Schubert calculus. In Combinatoire et Représentation du Groupe Symétrique, Lecture Notes in Math., no. 579, pp. 217-51. Berlin: Springer.

Stanley, R.P. (1982). Some aspects of groups acting on finite posets. J. Combinatorial Theory (A), 32, 132-161.

Weyl, H. (1946). The Classical Groups. Princeton, N.J.: Princeton University Press.

IRREGULARITIES OF PARTITIONS

Ramsey theory, uniform distribution

V.T.Sós
Eötvös Loránd University, Budapest, Múzeum krt. 6-8, Hungary

Abstract. In this survey we discuss a common feature of some classical and recent results in number theory, graph theory, etc. We try to point out the fascinating relationship between the theory of uniformly distributed sequences and Ramsey theory by formulating the main results in both fields as statements about certain irregularities of partitions. Our approach leads to some new problems as well.

INTRODUCTION

In 1916 Hermann Weyl published his classical paper entitled "Über die Gleichverteilung von Zahlen mod Eins". This was intended to furnish a deeper understanding of the results in diophantine approximation and to generalize some basic results in this field. The theory of uniformly distributed sequences has originated with this paper. In the last decades this subject has developed into an elaborate theory related to number theory, geometry, probability theory, ergodic theory, etc.

Curiously enough, Issai Schur's paper entitled "Über die Kongruenz $x^n+y^n \equiv z^n$ (mod p) " appeared in the very same year. He proved that if the positive integers are finitely colored, then there exist x, y, z having the same color so that $x+y=z$. Though Ramsey theory has various germs, Schur's theorem can be regarded as the first Ramsey-type theorem. Now literally the same applies to Ramsey theory as to the theory of uniform distribution:

In the last decades Ramsey theory became an elaborate theory related to number theory, geometry, probability theory, ergodic theory, etc.

It took about half a century for both fields to become coherent theories. It took more than a decade to realize the close relationship between the two seemingly unrelated areas. The interaction between the theory of uniform distribution and combinatorics in general is indicated in several works. We list without claiming completeness some of them: Erdős & Spencer (1972), Graham, Rothschild & Spencer (1980), Olson & Spencer (1978), Niederreiter (1972), Tijdeman (1980). A breakthrough in this direction has been achieved in a recent series of papers by J. Beck (1981 a, b, c, d), (1983 a, b, c, d).

1 *FORMULATION OF THE GENERAL PROBLEM*

As introductory examples we consider some classical theorems. The first is one of the basic results in the theory of uniform distribution.

Let $x_1,\dots,x_N$ be N points in the unit square. Let I be an aligned rectangle, i.e. one with sides parallel to those of the unit square. Denote by $Z(I)$ the number of points x_i, $1\le i\le N$ in I. $|I|$ denotes the Lebesgue-measure of I.

Theorem 1.1. *(Schmidt (1972)). There exists an aligned rectangle* I_o *such that*

$$|Z(I_o)-N|I_o|| > c\log N$$

holds, where $c>0$ *is a positive absolute constant.*

Theorem 1.2. *(Roth (1964)). Let* $[N]=\{1,\dots,N\}$. *For any partition* $[N]=S_1\cup S_2$, $S_1\cap S_2=\emptyset$ *there exists an arithmetic progression* $P=\{a,a+d,\dots,a+kd\}\subset[N]$ *such that*

$$||P\cap S_1|-|P\cap S_2|| > cN^{1/4}$$

holds, where c *is a positive absolute constant.*

Theorem 1.3. *(Ramsey (1930)). For* $n>n_o(t)$ *if the edges of* K_n *(the complete graph on* n *vertices) are 2-colored, then there must be a monochromatic* K_t.

The theorems stated above have a common feature. In all three of them we are given an underlying set S and a family of subsets of this set, and (in all three cases) the claim is that the underlying set has no partition which splits proportionally or equally each set contained in the given family.

We now give a formulation of the general problem.

Let S be a set and $A\subseteq 2^S$ a family of subsets of S. Let G denote the set of functions (partitions, colorings) $g: S\to\{1,\dots,r\}$.

Or more generally let us fix real numbers $\alpha_1,\ldots,\alpha_r$ and let G denote the set of functions $g : S \to \{\alpha_1,\ldots,\alpha_r\}$. We shall use various functions

$$\Delta : G \times A \to R$$

to measure the discrepancy (weighted nonuniformity) $\Delta_g(A)$ of $g \in G$ on A . Given such a discrepancy function, our goal is to estimate the following quantities:

I The discrepancy of g over A :

$$\Delta_g(A) = \sup_{A \in A} |\Delta_g(A)|$$

or e.g.

$$\Delta_g^p(A) = \Big(\sum_{A \in A} |\Delta_g(A)|^p\Big)^{\frac{1}{p}}$$

and the discrepancy of G over A :

$$\Delta_G(A) = \inf_{g \in G} \Delta_g(A) .$$

II The one-sided discrepancies of g over A :

$$\Delta_g^+(A) = \sup_{A \in A} \Delta_g(A)$$

$$\Delta_g^-(A) = |\inf_{A \in A} \Delta_g(A)|$$

and

$$\Delta_G^+(A) = \inf_{g \in G} \Delta_g^+(A) , \quad \Delta_G^-(A) = \inf_{g \in G} \Delta_g^-(A) .$$

In some cases we consider only partitions of S which satisfy certain requirements. This means that we have a $G^* \subseteq G$ and we investigate $\Delta_{G^*}(A) = \inf_{g \in G^*} \Delta_g(A)$.

Consider the case $r=2$. Here the simplest problem is the following:

We wish to find a two-coloring of S so that every set $A \in \mathcal{A}$ is partitioned by this into parts of size as equal as possible. Now a two-coloring of S can be given by a function $g: S \to \{+1,-1\}$. The function Δ defined by

$$\Delta_g(A) = \sum_{x \in A} g(x)$$

measures the discrepancy of this partition on A.

$$\Delta(\mathcal{A}) = \inf_g \sup_{A \in \mathcal{A}} \left| \sum_{x \in A} g(x) \right|$$

measures (in supremum norm) how well under the requirement above the set S can be partitioned.

Now consider a more general problem. We want to find a system of representatives of $\mathcal{A}$ so that in every set $A \in \mathcal{A}$ the number of representatives is proportional to $|A|$, i.e. for a given $\alpha \in (0,1)$ it is as close to $\alpha|A|$ as possible. This means that we consider partitions of ratio α, $1-\alpha$. To handle this problem now we take the functions $g: S \to \{\alpha-1, \alpha\}$ and

$$\Delta_g(A) = \sum_{x \in A} g(x) .$$

As above, define

$$\Delta(\mathcal{A}) = \inf_g \sup_{A \in \mathcal{A}} \left| \sum_{x \in A} g(x) \right| .$$

Note that if $g^{-1}(\alpha-1) = S_1$, $g^{-1}(\alpha) = S_2$, then

$$\sum_{x \in A} g(x) = |A|\alpha - |S_1 \cap A| .$$

Consequently, this measures the discrepancy in an appropriate way.

Let us reformulate the above theorems in this general setting.

From the proof of Schmidt's theorem it is easy to see that it can be given in the following finite form.

Theorem 1.1*. *(the finite version of Schmidt's theorem). Let*

$$S = \{(i,j);\ 1 \leq i \leq N\ ,\ \ 1 \leq j \leq N\}\ ,$$

$$A_{k,\ell} = \{(i,j)\ :\ 1 \leq i \leq k\ ,\ \ 1 \leq j \leq \ell\}$$

and

$$A = \{A_{k,\ell}\ :\ 1 \leq k \leq N\ ,\ \ 1 \leq \ell \leq N\}\ .$$

Set

$$G = \{g\ :\ S \to \{1 - \tfrac{1}{N}, -\tfrac{1}{N}\}\}$$

and $\Delta_g(A) = \sum_{x \in A} g(x)$.

Then

$$\Delta_G(A) \geq c\log N$$

where c *is a positive absolute constant.*

Further, Roth's theorem can be formulated in an obvious way as follows.

Theorem 1.2*. *(Roth). Let* $S=[N]$, $A = \{A\ :\ A \subseteq S\ ,\ A$ *is an arithmetic progression*$\}$, $G = \{g\ :\ N \to \{+1,-1\}\}$. *Put* $\Delta_g(A) = \left|\sum_{x \in A} g(x)\right|$. *Then*

$$\Delta_G(A) > cN^{1/4}$$

where c *is a positive absolute constant.*

Finally, a quantitative form of Ramsey's theorem says:

Theorem 1.3*. *Let* $|V|=n$, $S=[V]^2$, $A_k = \{A : A = [X]^2,\ X\ \ S$, $|X|=k\}$, $G = \{g\ :\ S \to \{+1,-1\}\}$. *Define* $\Delta_g(A) = \sum_{x \in A} g(x)$. *Then*

$$\Delta_G(A) = \binom{k}{2} \quad if \quad k < \frac{\log n}{2\log 2}\ .$$

2 *PARTITIONS OF THE SET OF INTEGERS AND LATTICE POINTS IN* R^n

We start by discussing results for arithmetic progressions.

(a) For arithmetic progressions we have both Ramsey type and discrepancy theorems, both expressing in a certain sense that we can not have too good partitions of the integers. In the case of Ramsey-type theorems this means that for any partition we must have a monochromatic arithmetic progression. The discrepancy-theorems mean that for any partition there exists an arithmetic progression, where one class has a large preponderance. So it should be clear that the seemingly qualitative difference between these statements is actually a quantitative one. If we focus on the short arithmetic progressions, the situation is so bad, that we must have an arithmetic progression where all but one colors are missing. If we focus on longer arithmetic progressions, this changes to a weaker preponderance phenomenon. This viewpoint helps to realize that we still have gaps in our knowledge. We will formulate the problems explicitly later.

(b) The methods used in the proofs of these results are good illustrative examples of the fascinating and fruitful relationship between the different fields. We will not give proofs, we just mention that e.g. in the discrepancy theorem 2.3 (below) for the lower bound Roth (1964) used a deep analytic method. Recently, an ingenious argument using only a combinatorial hypergraph theorem has been given by Beck (1981 a) which shows that Roth's lower bound is nearly sharp.

(c) In Ramsey theory van der Waerden's theorem for arithmetic progressions is widely generalized in different ways. However this is not the case with the discrepancy theorems. Having this common setting of the results on arithmetic progressions, we may again realize the gap, for quite a few general structures we know Ramsey-type generalizations but no discrepancy theorems.

Now we list a few theorems which may justify what we said above. Of course, here we can give just a sample of the results. First we mention the celebrated theorem of van der Waerden (1927):

Theorem 2.1. *(van der Waerden). If* $N > W(k,r)$ *, then for every* r *-coloring of* $[N]$ *at least one class must contain an arithmetic progression of length* k .

Though van der Waerden's theorem appears to be a Ramsey-type theorem, actually there is another reason for its validity. Erdős and

Turán conjectured already in 1936, that a density theorem is in the background. More exactly, they conjectured the following.

Let $r_k(n)$ denote the greatest integer r such that there exists a sequence

$$1 \leq a_1 < a_2 < \ldots < a_r \leq n$$

which does not contain an arithmetic progression of length k. Then $r_k(n) = o(n)$.

This was proved by Roth (1952) for $k=3$ and it was a great achievement when Szemerédi (1973) proved it for arbitrary k.

<u>Theorem</u> 2.2. *(Szemerédi). For every* $k \geq 3$

$$r_k(n) = o(n) .$$

Remark. No good estimates for $W(k,r)$ resp. for $r_k(n)$ are known. A lower bound is

$$W(k,2) > \frac{2^k}{k} .$$

However, the best known upper bound grows faster than

$$\left. 2^{2^{\cdot^{\cdot^{\cdot^{2}}}}} \right\} t$$

for any t. For reference see Erdős & Graham (1980).

Roth (1964) started to study the discrepancy problems for arithmetic progressions.

Let $\mathcal{P}$ denote the set of arithmetic progressions in $[N]$ and $g : [N] \to \{+1,-1\}$ a two-coloring of $[N]$. Let $P \in \mathcal{P}$. Set

$$\Delta_g(P) = \sum_{x \in P} g(x) ,$$

$$\Delta_g(P) = \max_{P \in \mathcal{P}} |\Delta_g(P)|$$

and

$$\Delta(P;N) = \min_g |\Delta_g(P;N)| .$$

In the next theorems c is always a positive absolute constant.

Theorem 2.3. *(Roth (1964)). For any two-coloring* $g : [N] \to \{+1,-1\}$

$$\Delta_g(P) > cN^{1/4} .$$

Roth conjectured that for every $g : [N] \to \{+1,-1\}$ $\Delta_g(P) > c\sqrt{N}$. This was disproved by Sárközy (1972). Recently Beck (1981 c) proved by an ingenious combinatorial argument that Roth's lower bound is nearly sharp. (See the combinatorial lemma in § 4.)

Theorem 2.4. *(Beck).*

$$\Delta(P;N) < cN^{1/4} (\log N)^{5/2}.$$

There are different variations of the discrepancy theorems for arithmetic progressions. Actually Roth proved a more general theorem, which shows that the discrepancy of arithmetic progressions is large on average. A quantitative form of this is given in the following

Theorem 2.5. *(Roth (1964)). Let*

$$P_{h,q}(m) = \{1 \le a \le m ; \quad a \equiv h \pmod q\} .$$

Let $A \subseteq [N]$ *be fixed and let*

$$\Delta_{h,q}(m;A) = \Delta_{h,q}(m) = |(A \cap P_{h,q}(m)| - \frac{|A|}{|N|} P_{h,q}(m)|| .$$

Set

$$V_q(m) = \sum_{k=1}^{q} \Delta_{h,q}^2(m) .$$

Then for any integer Q

$$\sum_{q=1}^{Q} q^{-1} \sum_{m=1}^{N} V_q(m) + Q \sum_{q=1}^{Q} V_q(N) \gg \frac{|A|}{|N|}(1 - \frac{|A|}{|N|}) Q^2 N .$$

Quoting Roth, this theorem says that a sequence A cannot be well distributed simultaneously among and within all congruence classes. (The choice $Q=N^{1/2}$ yields Theorem 2.3.)

In Roth's theorem the lower bound for the discrepancy depends on the ratio of the partition. But we do not have any information about the difference d of the arithmetic progressions of large discrepancy. Now we list a few theorems concerning this problem.

Theorem 2.6. *(Roth (1964)). Given any* $g : N \to \{+1,-1\}$ *for every integer* k *there is an arithmetic progression* P *of difference* $d > k$ *such that*

$$\Delta_g(P) > c\sqrt{d} .$$

Beck & Spencer (1983), using purely combinatorial arguments, proved that Roth's lower bound is nearly sharp. One of their results which gives upper bound for $\Delta(P)$ depending on the difference d is the following.

Theorem 2.7. *(Beck & Spencer (1983)). Let* $\varepsilon > 0$ *be arbitrary. Given any* $n > n_0(\varepsilon)$ *there is a two-coloring* $g : N \to \{+1,-1\}$ *such that for any arithmetic progression* P *of difference* $n^{\varepsilon} < d < n$ *and of arbitrary length*

$$\Delta_g(P) < c\sqrt{d}\,(\log n)^{7/2} .$$

Problem. Let P_t denote the set of arithmetic progressions of length t. By Roth's theorem we know that given any two-coloring $g : [N] \to \{+1,-1\}$, there must be an arithmetic progression with discrepancy $\Delta_g(P) > cN^{1/4}$ and hence of length $t > cN^{1/4}$.

Problem. What happens if $c \log N < t < N^{\varepsilon}$? Find upper and lower bounds for $\Delta(P_t)$.

The first generalization of van der Waerden theorem was given by Gallai (1931). Let $A=\{v_1,\dots,v_k\}$ be a subset of R^m. $B=\{x_1,\dots,x_k\}$ is *homothetic* to A if, under a suitable ordering of B, there exist $\lambda \in R$, $\lambda \neq 0$ and an $a \in R^m$ so that

$$x_i = \lambda v_i + a , \quad 1 \leq i \leq k .$$

<u>Theorem</u>. *(Gallai). Let* A *be an arbitrary finite subset of* R^m *. Given any* r *-coloring of* R^m *there exists a monochromatic* $B \subset R^m$ *homothetic to* A .

A further generalization is the Hales & Jewett (1963) theorem, which is considered as one of the most basic theorems in Ramsey-theory. (See Graham & al. (1980).)

Definition. Let

$$C_t^n = \{(x_1,\ldots,x_n) \; ; \; x_i \in \{0,\ldots,t-1\}\} \; .$$

A *line* in C_t^n is a set of suitably ordered points $x_0,\ldots,x_{t-1}$, $x_i = (x_{i1},\ldots,x_{in})$ so that for each coordinate j , $1 \le j \le n$ either

$$x_{oj} = x_{1j} = \ldots = x_{t-1,j}$$

or

$$x_{\ell j} = \ell \qquad \text{for} \quad 0 \le \ell < t$$

and the latter holds for at least one j .

<u>Theorem</u>. *(Hales & Jewett (1963)). For every* r , t *there exists a least integer* $HJ(r,t)$ *so that, for* $N > HJ(r,t)$ *if the vertices of* C_t^N *are* r *-colored, then there exists a monochromatic line.*

This is a corollary of a much more general Ramsey theorem (Graham-Leeb-Rothschild (1972) which also implies the following theorem.

Let A be an arbitrary finite field and let A^n be the n -dimensional space over A . For every r , t , k positive integers there exists an $N_t(k;r)$ so that if the t -dimensional linear subspaces of A^r are r -colored then there exists a k -dimensional vector space all of whose t -dimensional linear subspaces have the same color.

Remark. It is easy to see that van der Waerden's theorem is also a corollary of the Hales-Jewett theorem. For this consider the integers a , $0 \le a < t^N$ and the base- t -representation.

$$a = \sum_{i=0}^{N} a_i t^i \; , \qquad 0 \le a_i < t \; .$$

An r-coloring of $[t^N]$ induces an r-coloring of C_t^N. This has a monochromatic line, if N is large enough. But it is easy to see that a monochromatic line corresponds to a monochromatic arithmetic progression of length t.

Problem. We have the discrepancy-results for arithmetic progressions. Is it possible to get discrepancy-results for C_t^n ?

A common generalization of van der Waerden's and Schur's theorem was given by R. Rado (1933 a,b).

Let C be an $m\times n$ matrix of integer entries, $C\underline{x} = 0$ a system of homogeneous linear equations in the variables $x_1,\ldots,x_n$. We say that C is regular if given any finite coloring N, there exists $\underline{x} = (x_1,\ldots,x_n)$ such that $C\underline{x} = 0$ and $x_1,\ldots,x_n$ are the same color.

We say that the matrix C satisfies the column-condition if after a suitable rearranging of the column-vectors $c_1,\ldots,c_n$ one can find $1 \le k_1 < \ldots < k_t = n$ such that for

$$A_i = \sum_{j=k_{i-1}+1}^{k_i} c_j$$

we have $A_1=0$ and for $1 < i \le t$, A_i is a linear combination of $\underline{c}_1,\ldots$ $\ldots,c_{k_i-1}$.

Theorem. *(Rado). C is regular on N iff C satisfies the column-condition.*

Remark. It is easy to see that Schur's theorem is a consequence of Rado's theorem. Namely, if we have a single equation

$$a_1x_1 + \ldots + a_nx_n = 0 ,$$

the column-condition means that some nonempty subset of the c_i sums to zero. Evidently $x+y-z=0$ satisfies this condition.

To obtain van der Waerden's theorem as a special case we consider the system

$$\begin{aligned} x_2 &= x_1 + d \\ &\vdots \\ x_k &= x_{k-1} + d . \end{aligned}$$

Similarly as in the case of the Hales & Jewett theorem Rado's theorem is not accompanied by general discrepancy theorems.

3 *PARTITIONS ON GRAPHS*

The results here belong partly to Ramsey theory, partly to extremal graph theory. As there are excellent monographs on both subjects (Graham, Rothschild & Spencer (1980), Bollobás (1980)), we will mention only those basic results which are relevant for our present aim. However we will give a more detailed discussion of results which are more recent ones and are not contained in the books mentioned above.

First we consider partitions of $[n]^{\ell}$, with respect to the family of complete graphs.

Most of the results refer to the case $\ell=2$. Still, we formulate Ramsey's theorem for arbitrary ℓ.

Theorem 3.1. *(Ramsey (1930)). For all integers* ℓ, r, $k_1,\ldots,k_r$ *there exists a minimal integer* $R_{\ell}(k_1,\ldots,k_r)$ *with the following property:*

if $n > R_{\ell}(k_1,\ldots,k_r)$, *given any* r*-coloring* $g : [n]^{\ell} \to \{1,\ldots,r\}$, *there exist an* i, $1 \leq i \leq r$ *and a set* $S \subset [n]$ *such that* $|S|=k_i$ *and* $[S]^{\ell}$ *is colored* i.

Little is known about the Ramsey-numbers $R_{\ell}(k_1,\ldots,k_{\ell})$.

Here we mention just two results for the simplest cases, which are relevant in our discussion:

$$(*) \qquad 2^{k/2} < R_2(k,k) < 4^k$$

$$(**) \qquad ck^2(\log k)^{-2} < R_2(3,k) < ck^2(\log k)^{-1} ,$$

(see Graham, Rothschild & Spencer (1980)).

There are several interpretations of Ramsey's theorem. Graham, Rothschild & Spencer (1980) give a deep analysis of this question. We repeat only their quotation of Burkill & Mirsky (1973). "There are numerous theorems in mathematics which assert, roughly speaking, that every system of certain class possesses a large subsystem with a higher degree of organization than the original system." We emphasize here another aspect of Ramsey's theorem.

Suppose we want to two-color the edges of K_n by red and blue so that in every K_k we have about the same number of red and blue edges. Now the situation is similar to the case of arithmetic progressions. By (*) we know that if $k < \frac{1}{2\log 2}\log n$, the situation is so bad that for any two-coloring we must have a monochromatic K_k. As k gets larger, we can two-color more uniformly, with respect to the K_k's, though we still have the preponderance phenomenon.

A quantitative form of this is given by the following theorem.

<u>Theorem</u> 3.2. *(Erdős-Spencer (1972)). Let* $g : [n]^2 \to \{+1,-1\}$ *and*

$$\Delta_g(n;t) = \max_{\substack{S \subset [n] \\ |S| \le t}} \Big|\sum_{x \in [S]^2} g(x)\Big| .$$

Define

$$\Delta(n;t) = \min_g \Delta_g(n;t) .$$

Then

$$\Delta(n;t) = \binom{t}{2}, \quad \textit{if} \quad t \le \frac{\log n}{2\log 2},$$

and

$$10^{-3}\, t^{3/2}\sqrt{\log \frac{5n}{t}} \le \Delta(n;t) \le t^{3/2}\sqrt{\log\frac{5n}{t}}, \quad \textit{if} \quad t > \frac{2\log n}{\log 2} .$$

Corollary. Let

$$\Delta(n) = \min_g \max_{S \subset [n]} \Big|\sum_{x \in [S]^2} g(x)\Big| .$$

Then

$$c_1 n^{3/2} < \Delta(n) < c_2 n^{3/2}$$

where c_1, c_2 are positive absolute constants.

This theorem has a generalization for hypergraphs.

Theorem 3.4. *(Erdős & Spencer). Let* $g : [n]^k \to \{+1,-1\}$. *Set*

$$\Delta(n;t) = \min_{g} \max_{\substack{S \subset [n] \\ |S| \leq t}} \Big| \sum_{x \in [S]^k} g(x) \Big| .$$

Then

$$c_1\binom{t}{k} \leq \Delta(n;t) \leq \binom{t}{k} \quad \text{if} \quad t \leq (\log n)^{\frac{1}{k-1}}$$

and

$$c_2\, t^{\frac{k+1}{2}} \sqrt{\log \frac{5n}{t}} \leq \Delta(n;t) \leq t^{\frac{k+1}{2}} \sqrt{\log \frac{5n}{t}}$$

where c_1, c_2 *are positive absolute constants.*

The case $k_1=\ldots=k_\ell$ of Ramsey theorem is called "symmetric case", otherwise it is called "asymmetric".

We know that - concerning the theorems above - the best colorings are the random ones. Here "best" means that the largest monochromatic complete graph is as small as possible, resp. the discrepancy in the K_t 's is as small as possible. In the symmetric case if $r=2$ this means that in the best colorings each color class contains about the half of the edges.

Hence if we consider partitions of given ratio, α , $1-\alpha$ and $\alpha \neq \frac{1}{2}$, the same discrepancy phenomenon will appear, but the quantitative results will be different.

As to the asymmetric case, here we suppose $k = o(\ell)$ and $n \leq R(k,\ell)-1$. Let us consider the two-colorings of K_n which contain neither a red K_k nor a blue K_ℓ . It is easy to see that if ℓ is much larger than k , then the number of red edges in such a coloring will be much smaller than the number of blue edges. The ratio tends to 0 as $\frac{k}{\ell} \to 0$. The conjecture is that for fixed k , $R(k,\ell) \approx \ell^{k-1+o(1)}$.

All these facts make plausible that if we have a restriction on the number of edges in the graph (on the ratio of the partition) this will increase the size of the complete graphs we can ensure.

Another aspect of the results is the following:

If we make a comparison between van der Waerden's theorem and Roth's theorem for arithmetic progressions on the one hand and Ramsey's theorem, Erdős & Spencer theorem for complete graphs on the other hand, the similarity is clear.

In both cases we have a set S and a family $A \subseteq 2^S$. For any two-coloring of S among the small sets in A there must be a monochromatic one, and among the larger sets still there must be one in which there is a certain discrepancy.

However, the background of the results for the two structures, at least for the monochromatic case is different.

For arithmetic progressions actually a density theorem yields the result. Namely Szemerédi's theorem means that for every $c > 0$, if $n > n_0(k)$, $A \subseteq [n]$, and $|A| > cn$, then A must contain an arithmetic progression of length k. Since for every two-coloring at least one color-class contains more than $\frac{n}{2}$ elements, we must have an arithmetic progression of length k in it.

Although we have a density theorem for complete graphs, (Turán's theorem) we need more than half of the edges to ensure the existence of a complete K_k. Therefore, if we consider the two-colorings of the edges of K_n, just a density argument will not be enough to ensure the existence of a monochromatic K_k. This is the reason why we have a new class of problems and results for graphs.

Let H^k be an ℓ-uniform hypergraph. Let $f(n;H^k)$ be the minimal integer e such that every k-uniform hypergraph on n vertices and more than e edges contains a subgraph isomorphic to H^k. A $G_n^k(V;E)$ is called an extremal graph belonging to H^k, if $|V|=n$, $|E|=f(n;H^k)$ and G_n^k does not contain sugraphs isomorphic to H^k.

The determination (or estimation) of $f(n;H^k)$ is the fundamental problem of extremal graph theory started by Turán (1941).

As to the density theorems for graphs, we mention only results which are relevant here.

First we consider the case $\ell=2$ and $H^2=K_k$.

Theorem 3.5. *(Turán (1941)). Let* $n \equiv r \bmod (k-1)$, $0 \leq r < k-1$. *Then*

$$f(n;K_k) = \frac{1}{2}\,\frac{k-2}{k-1}\,(n^2-r^2)+\binom{r}{2} .$$

There is a unique extremal graph, the complete $(k-1)$*-partite graph having* $[\frac{n}{k-1}]$ *resp.* $[\frac{n}{k-1}]+1$ *vertices in each class.*

Observe that

(1) as we said above, even in the case $k=3$ we need more than half of the edges to ensure the existence of a K_k.

(2) From our point of view it is important that the extremal graph contains a very large independent set of size $[\frac{n}{k-1}]$.

These indicate, too, that if we have a restriction on the number of edges in one color-class (on the ratio of the partition), then for the asymmetric case this will change the size of the complete graphs we can ensure in one color-class.

To formulate a slightly more general problem consider partitions of $[n]^2$ of ratio α, $1-\alpha$. For which pairs (k,ℓ) is it true, that either the first class contains a K_k or the second class contains a K_ℓ. (Evidently, by Ramsey theorem, with $n\to\infty$ $\max(k,\ell)\to\infty$.)

Or more generally, let $RT(n;k,\ell)$ be the set of integers e, for which there is a coloring of $[n]^2$ such that the number of red edges is e and neither a red K_k nor a blue K_ℓ exists. What can be said about $RT(n;k,\ell)$? In particular, we are interested in the value of

$$\max RT(n;k,\ell) \quad \text{and} \quad \min RT(n;k,\ell) .$$

Results which give information on this question are called *Ramsey-Turán type theorems*.

Since we do not have too much knowledge about the Ramsey numbers $R(k,\ell)$ one can expect that the most we can have are asymptotic results.

We start with a result of Erdős and Szemerédi on the symmetric case.

Theorem 3.6. *(Erdős & Szemerédi (1972)). Let* $r\geqq 2$. *Let* $G(V;E)$ *be a graph with* $|V|=n$, $|E|\leqq\frac{1}{r}\binom{n}{2}$. *There exists a positive absolute constant* c *so that either* G *or its complement contains a* K_k *with* $k > c\,\frac{r}{\log r}\log n$.

Remark. In the present setting the above theorem can be formulated as follows. Let

$$G^* = \{g \mid g:[n]^2\to\{+1,-1\},\ \sum_{x\in[n]^2} g(x) = (1-\tfrac{2}{r})\tbinom{n}{2}\} .$$

Set

$$\Delta_g(n;t) = \max_{\substack{S\ [n] \\ |S|=t}} \left| \sum_{x \in [S]^2} g(x) \right|$$

and

$$\Delta^{(r)}(n;t) = \min_{g \in G^*} \Delta_g(n;t) .$$

Then

$$\Delta^{(r)}(n;t) = \binom{t}{2} \quad \text{if} \quad t < c \frac{r}{\operatorname{Log} r} \log n .$$

Problem. Find a common generalization of Theorem 4.2 and Theorem 4.3: give estimates for $\Delta^r(n;t)$ if $t > c \frac{r}{\operatorname{Log} r} \log n$.

Remark 2. The Erdős-Szemerédi theorem gives a partial answer to the following question:

Let $f(n;r)$ be the largest integer k such that for any r-coloring of $[n]^2$ there exists an $S \subset [n]$, $|S| = k$ for which $[S]^2$ meets at most $r-1$ color-classes. Since at least one color-class contains not more than $\frac{1}{r}\binom{n}{2}$ edges, by the Erdős-Spencer theorem $f(n;r) >$ $> c \frac{r}{\operatorname{Log} r} \log n$.

Problem. Let $1 \leq s < r$. Let $f(n;r,s)$ denote the largest integer k such that for any r-coloring of $[n]^2$ there exists an $S \subset [n]$, $|S| = k$, for which $[S]^2$ meets at most s color-classes. Find upper and lower bounds for $f(n;r,s)$.

Let

$$RT(n;k,\ell) = \max RT(n;k,\ell) .$$

As to the function $RT(n;k,\ell)$ in the asymmetric case, most of the results are asymptotic estimates for the case when ℓ is replaced by a function of n which is $o(n)$. For this we will use the symbol $RT(n;k,o(n))$.

For $t \geq 3$ put

$$a_t = \frac{1}{2}\frac{t-3}{t-1} = \frac{1}{2}\frac{3t-9}{3t-3} \quad \text{if } t \text{ is odd}$$

$$a_t = \frac{1}{2}\frac{3t-10}{3t-4} \quad \text{if } t \text{ is even.}$$

(The sequence $0 = a_3, a_4, \ldots = 0, \frac{1}{8}, \frac{1}{4}, \frac{2}{7}, \ldots$ is strictly increasing.)

Theorem 3.7. *(Erdős-Hajnal-T.Sós-Szemerédi (1983)). For* $t \geq 3$

$$RT(n;t,o(n)) = a_t\, n^2(1+o(1))\ .$$

The investigation of $RT(n;t,o(n))$ started in T. Sós (1969) and Erdős - T. Sós (1969). The above result for t odd was proved in Erdős & T. Sós (1969). The case when t is even turned out to be much harder. It was proved only much later in Szemerédi (1973) and in Bollobás & Erdős (1976) that $RT(n;4,o(n)) = \frac{n^2}{8}(1+o(1))$. (Szemerédi (1973) gives the upper bound and Bollobás & Erdős (1976) the construction yielding the lower bound.

Results for an other range of the parameters k, ℓ , or in other words, for min $RT(n;k,o(n))$ are given in Ajtai-Komlós-Szemerédi (1981) , Komlós-Pintz-Szemerédi (1982), Ajtai-Erdős-Komlós-Szemerédi (1981), Ajtai-Komlós-Pintz-Spencer-Szemerédi (1983).

Theorem 3.8. *(Ajtai-Komlós-Szemerédi (1981)). Let* $G(V;E)$ *be a graph of* n *vertices,* e *edges. Let* t *denote the average degree:* $t = \frac{2e}{n}$ *and* $\alpha(G)$ *denote the maximum size of an independent set of vertices (independence or stability number).*

If G *does not contain a triangle, then*

$$\alpha(t) > c\frac{n}{t}\log t$$

(where c *is a positive absolute constant).*

The result is best possible as to the order of magnitude.

Remark. Without the assumption that G is triangle-free, by a simply greedy-algorithm argument only $\alpha(t) > c\frac{n}{t}$ would follow.

Theorem 3.9. *(Komlós-Pintz-Szemerédi (1982)). Let* $H^3(V;E)$ *be a* 3 *-uniform hypergraph of* n *vertices and* e *edges. Let* t *denote the average degree;* $t = \frac{3e}{n}$ *and let* $\alpha(G)$ *denote the maximum size of an independent set.*

If $H^3(V;E)$ *does not contain cycles of length* ≤ 4 *, (in a precise meaning, which we do not give here)*

$$\alpha(G) > c\frac{n}{\sqrt{t}}\log t .$$

(Here c *is a positive absolute constant.)*

See also Erdős-Komlós-Pintz-Spencer-Szemerédi (1983).

Remark. It is worth mentioning that the Ramsey-Turán type theorems were considered because of different applications. There is a sequence of papers by Erdős-Meir-T. Sós-Turán (1971), (1972 a), (1972 b), where Turán's theorem is employed to obtain results for the distribution of distances in metric spaces. Ajtai et al. (1981) applied their theorem in the investigation of Sidon-sequences. The result in Komlós et al. (1982) was the key lemma to disprove a more than 20 year-old conjecture of Heilbronn for the minimum area of triangles determined by n points in the unit square.

Remark. Observe that $RT(n;k,o(n)) \approx c_k\, f(n;k)$ with a $c_k < 1$. Surprisingly enough the situation for ℓ -uniform hypergraphs is different if $\ell \geq 3$. In Erdős & T. Sós (1982) it is proved that

$$RT^{(\ell)}(n;k,o(n)) \approx f^{(\ell)}(n;k)$$

where $RT^{(\ell)}(n;k,o(n))$ and $f^{(\ell)}(n;k)$ has a similar meaning for $\ell \geq 3$ as $RT(n;k,o(n))$ and $f(n;k)$ for $\ell=2$.

Now let us consider the discrepancy in colorings of the edges of K_n with respect to general graphs.

Let the graphs $G_1,\dots,G_r$ be fixed. Evidently it follows from Ramsey theorem that for n large enough, for every r -coloring of $[n]^2$ there exists i , $1 \leq i \leq r$ so that a copy of G_i occurs in the i th color. The problem is to find the least integer n for which this holds. Let $R(G_1,\ldots,G_r)$ denote this least integer. The investigation of $R(G_1,\ldots,G_r)$ was started in Gerencsér & Gyárfás (1967) where they considered the case $r=2$, $G_1=P_k$, $G_2=P_\ell$ (the paths of length k resp. ℓ).

Though the problem is more general, for some special class of graphs it is easier to get good estimates or even to get the exact values of $R(G_1,\ldots,G_r)$ than for $R(k_1,\ldots,k_r)$.

The Turán type results for arbitrary graphs are also relevant here. For all graphs H^2 with chromatic number $\chi(H^2)>2$ the asymptotic value of $f(n;H^2)$ is known.

Theorem 3.10. *(Erdős-Simonovits (1966)). Let* $\chi(H^2)=k$

$$f(n;H^2) \approx f(n;k) = \frac{1}{2}\frac{k-2}{k-1}n^2 + o(n^2) .$$

Even more is true.

Theorem 3.11. *(Erdős (1967), Simonovits (1968)). If* $k>2$, *the extremal graphs (having* $f(n;H^2)$ *edges without containing subgraphs isomorphic to* H^2 *) can be made isomorphic by adding to and deleting from* $o(n^2)$ *edges the Turán-graph (the complete* $(k-1)$ *-partite graph having* $[\frac{n}{k-1}]$ *or* $[\frac{n}{k-1}]+1$ *vertices each class).*

Remark. If $k=2$, the above theorem gives only that

$$f(n;\, H^2) = o(n^2) .$$

For most bipartite graphs the exact value of $f(n;H^2)$ or even asymptotic formula for it is not known, and to determine it is among the most difficult problems in extremal graph theory.

Roughly speaking, the maximum number of edges a graph may have without containing H^2 as a subgraph, asymptotically depends *only* on the chromatic number of H^2.

Why are the above results relevant in the problems for $RT(G_1,G_2)$? If $\chi(G_1)=\chi(G_2)=2$, $f(n;G_i)=o(n^2)$ $(i=1,2)$. Hence a density theorem ensures the existence of a monochromatic G_1 or G_2. This means that the situation in this case is similar to the case of arithmetic progressions.

Remark. Here we discussed the following type of problems:

(1) Ramsey-type problems;

(2) Turán-type problems;

(3) Ramsey-Turán-type problems;

(4) discrepancy-problems.

In all the four cases we considered the complete graphs as subgraphs. In cases (1) - (2) there are many results for other graphs too. In (2) we know that the chromatic number of a graph is the most relevant parameter which determines the behavior of $f(n;G)$.

In (1) most of the results are for graphs G_1 , G_2 when $\min_i \chi(G_i) = 2$ (see Graham-Rothschild-Spencer (1980), and in the general case we do not know which parameters of the graphs G_1, G_2 determine the Ramsey-function $R(G_1, G_2)$.

For (3) in Erdős-Hajnal-T. Sós-Szemerédi (1983) the function $RT(n;H,o(n))$ is defined as the maximal e for which there exists a graph G with n vertices and e edges, such that $H \not\subseteq G$ and the stability number of G is $o(n)$. Not even asymptotic results for $RT(n;H,o(n))$ are known in the general case. The lower and upper bounds proved in Erdős-Hajnal-T. Sós-Szemerédi show that here the arboricity number of the graph H is relevant.

In (4) only the complete graphs were considered. It would be interesting to have discrepancy theorems also for other graphs.

4 *PARTITIONS IN* R^n . *CLASSICAL THEORY OF UNIFORMLY DISTRIBUTED SEQUENCES*

To begin with the history we have to go back to the seventeenth century when Huygens wanted to give a mechanical model for the solar system using a system of gears. Each gear represented a planet.

The number of teeth on the gears had to be chosen so that the ratio of these numbers were close to that of the periods of revolution of the represented planets. At the same time the number of teeth on each gear was limited. So the mathematical problem was the following:

Given a real number α and N, find integers $0 \le p$, $q \le N$ so that $|\alpha - \frac{p}{q}|$ is as small as possible.

This was one of the germs of the theory of diophantine approximation. The theory developed much later, mainly due to the works of Ostrovski, Hecke, Hardy, Littlewood etc. It became clear that the approximability property of α depends on the partial quotients (a_k digits) of its continued fraction expansion $\alpha = \cfrac{1}{a_1 + \cfrac{1}{a_2 + \dots}}$ (denoted by $\alpha = [a_1, a_2, \dots]$). It became also clear that the approximability property of α is closely related to the distribution of the sequence $(\{n\alpha\})$ in $(0,1)$. ($\{n\alpha\}$ denotes the fractional part of α). Evidently, for every irrational α, the sequence $\{n\alpha\}$ is everywhere dense in $(0,1)$. The fact that it is uniformly distributed, expresses a stronger property.

Let E^k denote the k-dimensional unit cube $[0,1]^k$. $I = \mathop{X}\limits_{i=i}^{k} [a_i, b_i]$ be a box in E^k, $I(\underline{x}) = \mathop{X}\limits_{i=1}^{k} [0, x_i]$, $|I|$ be the Lebesgue measure of I.

Let $\omega = (u_n)$ be a sequence in E^k. We write $Z_\omega(N;I)$ (or simply $Z(N;I)$) for the number of $u_i \in I$, $1 \le i \le N$.

<u>Definition</u> 1. The sequence (u_n) is uniformly distributed in E^k if for every box $I \subset E^k$

$$\lim_{N\to\infty} \frac{Z_\omega(N;I)}{N} = |I| \tag{4.1}$$

holds.

An equivalent definition is the following.

<u>Definition</u> 2. Let $R(E^k)$ denote the set of Riemann-integrable functions on E^k. The sequence (u_n) is uniformly distributed in E^k if for every $f \in R(E^k)$

$$\lim_{N\to\infty} \frac{1}{N} \sum_{n=1}^{N} f(u_n) = \int_{E^k} f(x)dx \, . \tag{4.2}$$

The second definition seems to be less natural, however it is a more fruitful one. It indicates why uniformly distributed sequences are important in the theory of numerical integration (see Remark 4.1).

Further observe that we obtain an equivalent definition if we assume that (4.2) holds for a dense subset of $R(E^k)$. This indicates how the concept of uniformly distributed sequences can be generalized to topological groups.

For the general theory of uniformly distributed sequences see the excellent book of Kuipers-Niederreiter (1974).

Put $$\Delta_N^{\omega}(I) = |Z(N;I) - N|I||$$

and

$$\Delta_N^{\omega} = ||\Delta_N^{\omega}||_{\infty} = \sup_{I \subset E^k} \Delta_N^{\omega}(I) \ ;$$

$$||\Delta_N||_p = \left(\int_{E^k} |\Delta_N^{\omega}(I)(x))|^p \, dx\right)^{\frac{1}{p}} .$$

Δ_N resp $||\Delta_N||p$ measures (in different norms) the discrepancy of the sequence $u_1,\ldots,u_N$, their behavior for $N\to\infty$ measures the irregularity of the distribution of the infinite sequence (u_n) .

In the quantitative theory of uniform distribution a central problem is the investigation of the order of magnitude of the discrepancy-functions $||\Delta_N||_p$, Δ_N .

It is easy to see that a sequence ω is uniformly distributed in E^k iff $\Delta_n^{\omega}=o(N)$. But how small can $o(N)$ be?

The quantitative theory of uniformly distributed sequences started with the following conjecture of van der Corput (1935 a).

For an arbitrary sequence in $[0,1)$, $\sup_N \Delta_N = \infty$. This means that no sequence can be "too evenly" distributed. This was proved by van Aardenne Ehrenfest (1945) who showed that for an arbitrary sequence (u_n) for infinitely many N

$$\Delta_N > c(\log\log N)(\log\log\log N)^{-1} .$$

Roth (1954) strengthened this result. He proved the following more general theorem.

<u>Theorem</u> 4.1. *(Roth (1954)). A. For an arbitrary infinite sequence* (u_n) *in* E^k *and for every* $N > N_0$

$$\max_{1\le n\le N} \|\Delta_n\|_2 > c_k (\log N)^{\frac{k}{2}} .$$

B. For N *arbitrary points* $u_1,\dots,u_N$ *in* E^k

$$\|\Delta_N\|_2 > c'_k \log^{\frac{k-1}{2}} N \quad \textit{if} \quad N > N_o .$$

(Here c_k , c'_k *are positive absolute constants.)*

For $k=2$ Davenport (1956) and for $k\geq 3$ Roth (1979), (1980) proved that (apart from a multiplicative constant) these results on $\|\Delta_N\|_2$ are sharp.

The theorem implies (in a precise quantitative form), that the irregularity of the distribution increases with the dimension.

The problem of finding bounds for the discrepancy in supremum norm is more difficult. Since $\Delta_N \geq \|\Delta_N\|_2$, the preceding results give some lower bounds on Δ_N . For infinite sequences sharp results are known only for $k=1$, for finite sequences for $k=2$.

<u>Theorem</u> 4.2. *(Schmidt (1972)). A. For arbitrary infinite sequence* (u_n) *in* $(0,1)$ *and for every* $N > N_o$

$$\max_{1\le n\le N} \Delta_n > c \log N .$$

B. For arbitrary N *points* $u_1,\dots,u_N$ *in* E^2

$$\Delta_N > c' \log N$$

where c , c' *are positive absolute constants.*

This result is best possible apart from the multiplicative constant. E.g. if $u_n = \{n\alpha\}$ where α is an irrational number of bounded partial quotients ($a_k \leq K$, $k=1,2,\dots$), then for every N , $\Delta_N < c_K \log N$. Similarly, for the N points $u_n = \{\{n\alpha\},\frac{n}{N}\}$, $1\leq n\leq N$ in E^2 $\Delta_N < c_K \log N$.

We have mentioned that uniformly distributed sequences play an important role in the theory of numerical integration. The first result on this is the following.

<u>Theorem</u>. *(Koksma inequality (1942/43)). Let* f *be a function*

on $[0,1)$ *of bounded variation* $V(f)$ *and* $u_1,\dots,u_N$ *be* N *given points in* $[0,1)$ *with discrepancy* Δ_N *. Then*

$$\left|\frac{1}{N}\sum_{n=1}^{N} f(u_n) - \int_0^1 f(t)dt\right| \le V(f) D_N \frac{1}{N} .$$

Koksma's inequality has various generalizations, to higher dimensions too. These theorems show that in computing integrals (especially in higher dimensions) well distributed sequences are of great importance. The smaller the discrepancy is, the better the approximation will be. This led to the criticism of Monte Carlo methods, where randomly generated sequences are used. Since a random sequence has discrepancy $\sqrt{N \log\log N}$, it is more advantageous to work with well-distributed deterministic sequences of discrepancy of a power of $\log N$ (Niederreiter (1978)).

Remark. By an observation of Roth part A and part B of Theorem 4.1 and similarly of Theorem 4.2 are equivalent. Roth's argument (for $k=1$) is the following.

(a) Let $u_1,\dots,u_N$ be a sequence of points in $[0,1)$. Consider a corresponding set $(u_i, \frac{i-1}{N})$, $1\le i\le N$ of N points in E^2 .

(b) Let (x_i,y_i) , $1\le i\le N$ be a sequence of points in E^2 , arranged so that $y_1\le y_2\le\dots\le y_N$. Take the corresponding sequence $x_1,\dots$ $\dots,x_N$ in $[0,1)$.

In both cases the discrepancies of the two sequences are the same up to a universal constant factor.

Theorem 4.2. can be formulated as an assertion on partitions of a special finite hypergraph. (See Theorem 1.1*.) We obtain this finite version, if instead of E^2 (as underlying set) we take the $N\times N$ lattice. The continuous or discrete formulation makes no difference as long as the members of the family A are the set of points in aligned rectangles. However, if we consider other families too, the situation changes. Firstly, it may depend on the underlying set, which families are worth considering. Secondly, it depends on the family which form fits better to the distribution problem or to the proof.

5 *GEOMETRICAL STRUCTURES*

In this section we discuss a variety of questions where the underlying set S is either the k-dimensional unit cube E^k , or (in the discrete version) the $N\times N$ $(N\times N\times\ldots\times N)$ lattice defined as the set of points with integer coordinates $1\leq k\leq N$, $1\leq \ell\leq N$. A is a family of simple geometrical objects, as aligned or tilted rectangles, triangles, balls, etc. (see Erdős (1964)).

The distribution (or partition) problems will be formulated in the following two forms.

(a) Given N points in S, how evenly can they be distributed with respect to the sets in A?

(b) Given N points in S, how "good" can a two-coloring of these points be with respect to the sets in A?

All the theorems here give quantitative results on the weaker preponderance phenomenon: none of them is Ramsey-type (ensuring monochromatic subsets). Of course, one could consider the Ramsey-type results for lattice points mentioned in § 2 as belonging to this group of problems, too (e.g. Roth, Hales & Jewett, Szemerédi's theorem, etc.).

Let $U_n=\{u_1,\ldots,u_N\}\subset S=E^2$, A be a family of subsets in R^k. Let $\mathbb{Z}(A;U_N)$ denote the number of points u_i, $1\leq i\leq N$ in $A\in A$. Set

$$\Delta(A;U_N) = |\mathbb{Z}(A;U_N) - N\mu(A\cap S)|$$

$$\Delta(A;U_N) = \sup_{A\in A} \Delta(A;U_N)$$

$$\Delta_N(A) = \inf_{U_N} \Delta(A;U_N) .$$

<u>Theorem</u> 5.1. *Let* $S=E^2$, A *be the family of right-angled triangles in* E^2 *with sides containing the right angle parallel to coordinate axes. Then for arbitrary* $\varepsilon>0$ *if* N *is large enough*

$$c_1 N^{1/4-\varepsilon} < \Delta_N(A) < N^{1/4}\sqrt{\log N} .$$

(For the lower bound see Schmidt (1969), for the upper bound Beck (1983 a).)

Remark. If A is the family of aligned rectangles, then

$$c_1 \log N < \Delta_N(A) < c_2 \log N .$$

(See Theorem 4.2.) Compare the two results. There exists a set of N

points in the unit square such that the discrepancy in every aligned rectangle is not larger than $c \log N$. However, there must exist an aligned rectangle such that splitting it into two right-angled triangles, both will have a discrepancy as large as $N^{\frac{1}{4}-\varepsilon}$ (but of course of different signs).

These results, and actually all the other ones below raise the problem on which properties of the family A does it depend whether the discrepancy is "large" or "small" (N^{α} or $(\log N)^{\beta}$). The situation is rather annoying, we do not have a complete understanding of the problem.

A result related to the above one is given in the following theorem.

Theorem 5.2. *Let* $S=E^2$, A *be the family of tilted rectangles in the plane (not necessarily contained in the unit square* E^2 *). Then*

$$c_1 N^{\frac{1}{4}-\varepsilon} < \Delta_N(A) < c_2 N^{\frac{1}{4}} \cdot \sqrt{\log N} .$$

The lower bound was proved by Schmidt (1969), the upper bound by Beck (1981 a).

A generalization for higher dimensions is the following theorem.

Theorem 5.3. *(Beck (1983)). Let* $U_N=\{u_1,\ldots,u_N\}$ *be a set of* N *points on the* k*-dimensional unit torus. Let* A *be the set of tilted squares with diameter at most* 1. *Then*

$$N^{\frac{1}{2}-\frac{1}{2k}}\sqrt{\log N} > \Delta_N(A) > c(d)N^{\frac{1}{2}-\frac{1}{2k}} .$$

Remark. Schmidt proved for $k=2,3$ the slightly weaker result that $\Delta_N(A) > N^{\frac{1}{2}-\frac{1}{2k}-\varepsilon}$.

Theorem 5.4. *(Schmidt (1969 b)).* $S=E^k$, A *be the family of balls in* R^k *(not necessarily contained in* E^k *). Then*

$$N^{\frac{1}{2}-\frac{1}{2k}} < \Delta_N(A) < c_1 N^{\frac{1}{2}-\frac{1}{2k}} \sqrt{\log N} .$$

Answering an old problem of Roth, very recently Beck proved the following theorem.

Theorem 5.5. *(Beck (1983)). Let* S *be the unit disc,* $\mathcal{B}$ *the set of halfplanes,* $\mathcal{A} = \{S \cap B \; ; \; B \in \mathcal{B}\}$. *Then*

$$c_1 N^{\frac{1}{4}-\varepsilon} (\log N)^{-1} < \Delta_N(\mathcal{A}) < N^{\frac{1}{4}+\varepsilon} \log N .$$

Theorem 5.6. *(Beck (1983c)). Let* (u_n) *be an infinite sequence of points in the plane. Then for every real number* r *there exists a tilted square* A_r *of side* r, *such that*

$$r^{\frac{1}{2}} \log r > |Z(A_r) - r^2| > c \cdot r^{\frac{1}{4}} .$$

(Here $Z(A_r)$ *denotes the number of points* $u_i \in A_r$ *.)*

This theorem is especially remarkable since this is the first one where a set of arbitrary size and of large discrepancy can be guaranteed.

For a set $U_N = \{u_1,\dots,u_N\} \subset S$ let

$$G = \{g \mid g : U_N \to \{+1,-1\}\}.$$

Set

$$D(\mathcal{A};U_N) = \min_g \sup_{A \in \mathcal{A}} \Big| \sum_{x \in A \cap U_N} g(x) \Big|$$

and

$$D_N(\mathcal{A}) = \sup_{U_N} D(\mathcal{A};U_N) .$$

G. Tusnady asked the order of magnitude of (or bounds for) $D_N(\mathcal{A})$ in the special case when $S=E^2$ and $\mathcal{A}$ is the family of aligned rectangles. Recently Beck investigated this problem, for other families $\mathcal{A}$ too.

Theorem 5.7. *(Beck (1981 b)). Let* $S=E^2$, $U_N \subset E^2$, $\mathcal{A}$ *be the family of aligned rectangles. Then*

$$c_1 \log N < D_N(A) < c_2(\log N)^4 .$$

Again, for tilted rectangle the discrepancy is much larger.

Theorem 5.8. *(Beck (1981 b)). Let* $S=E^2$, $U_n \subset E^2$, A *be the family of tilted rectangles. Then*

$$c_1 N^{\frac{1}{4}-\varepsilon} < D_N(A) < c_2 N^{\frac{1}{2}+\varepsilon} .$$

Theorem 5.9. *(Beck (1983 a)). Let* S *be the* $N\times N$ *lattice,* A *be the family of tilted rectangles in the* $N\times N$ *lattice. Then*

$$c_1 N^{\frac{1}{3}-\varepsilon} < D(A;S) < N^{\frac{1}{2}} \sqrt{\log N} .$$

If tilted rectangles not necessarily contained in the $N\times N$ square belong to A too, the discrepancy is even much larger.

Theorem 5.10. *(Beck (1983 c)). Let* S *be the* $N\times N$ *lattice,* A *be the family of tilted rectangles (not necessarily contained in the* $N\times N$ *lattice). Then*

$$c_1 N^{\frac{1}{2}} (\log N)^{-\frac{1}{2}} < D(A,S) < c_2 N^{\frac{1}{2}} (\log N)^{\frac{1}{2}} .$$

6 ($\{n\alpha\}$) -*SEQUENCES AND ERGODIC THEORY*

Most of the recent results for the distribution of point sets refer to the d -dimensional space for $d\geq 2$ and they have a definite geometric character. However, there is widely developed theory of uniformly distributed sequences in $[0,1)$ and only few of these results have been generalized for higher dimension.

We formulate all the results below for distributions of infinite sequences in $[0,1]$. According to the observation of Roth actually we could formulate the assertions for finite point-sets in E^2 as well.

The most important class of uniformly distributed sequences in $[0,1)$ is the class of sequences ($\{n\alpha\}$) for α irrational. These are the basic sequences in the theory of diophantine approximation. Further, these are the best "test-sequences": very often theorems which were found first for sequences $(\{n\alpha\})$ turned out to be true for more general ones. Finally we mention the relation of sequences $(\{n\alpha\})$ to topological transformations.

The discrepancy of $(\{n\alpha\})$ depends on the partial quotients a_k, $k=1,\dots$ of α. For every N and $x\in[0,1)$ there is an "explicit" formula for $\Delta_N([0,x))$ (see T. Sós (1974)). This leads e.g., to the following

Theorem 6.1. *Let* $\frac{p_k}{q_k}$ *be the* k*-th convergent of* α; $\frac{p_k}{q_k}=[a_1,\dots,a_{k-1}]$. *If* $q_k\leq N<q_{k+1}$ *then*

$$c_1\sum_{i=1}^{k} a_i < \max_{1\leq n\leq N}\Delta_n < c_2\sum_{i=1}^{k+1} a_i .$$

Consequently, if $a_i\leq K$, $i=1,\dots,$ *then*

$$\Delta_N < c_K \log N .$$

Much is known about the finer properties of the distribution. Though

$$\max_{1\leq n\leq N}\sup_I \Delta_n(I) > c\log N ,$$

there are intervals I in which the distribution is very good.

Theorem 6.2. *For the sequence* $(\{n\alpha\})$ *and for a fixed interval* I

$$\sup_N \Delta_N(I) < \infty ,$$

iff $|I|=\{k\alpha\}$ *for some integer* k.

The "if" part was proved by Hecke (1922) and much deeper "only if" part by Kesten (1966). Very elegant proofs and generalizations of this theorem in the framework of ergodic theory are due to Fürstenberg, Keynes & Shapiro (1973), Halász (1976), Petersen (1973).

On the other hand it is remarkable that this theorem (and further properties of Δ_N) has consequences for ergodic theory (see e.g.

Herman (1976 a,b), Deligne (1975).

Schmidt investigated the analogous question for arbitrary sequences in $[0,1)$.

Theorem 6.3. *(Schmidt (1974)). For an arbitrary sequence* (u_n) *in* $[0,1)$ *the lengths of all intervals* I *with* $\sup_N \Delta_N(I) < \infty$ *form at most a countable set.*

The ergodic theoretical generalization shows the essence of Kesten's theorem. Let (Ω,A,μ) be a probability space, $T:\Omega \to \Omega$ an ergodic transformation. For an $A \in A$ let $Z_N(A;x)$ denote the number of points $T^n x \in A$. $1 \le n \le N$. Set

$$\Delta_N(A;x) = Z_N(A;x) - N\mu(A) .$$

By Birkhoff's ergodic theorem, for fixed $A \in A$ for almost all $x \in \Omega$

$$\frac{1}{N} \Delta_N(A;x) \to 0 \text{ , if } N \to \infty .$$

Furstenberg, Keynes & Shapiro (1973), Petersen (1973), Halász (1976) proved the following generalization of Kesten's theorem:

Theorem 6.4. *If for* $A \in A$ $\sup_N |\Delta_N(A;x)|$ *is bounded on a set* $X \subset \Omega$ *of positive measure, then* $e^{2\pi i \mu(A)}$ *is an eigenvalue of* T *; i.e., there exists a function* $g \not\equiv 0$ *such that*

$$g(Tx) = e^{2\pi i \mu(A)} g(x) \quad \textit{for} \quad x \in \Omega .$$

On the other hand, for every eigenvalue $e^{2\pi i \mu}$ *there exists an* $A \in A$ *such that* $\mu(A) = \mu$ *and* $\Delta_N(A;x)$ *is bounded in* N *for almost all* $x \in \Omega$.

Remark. Kesten's theorem follows from the above one. To see this let $\Omega = R/\mathbb{Z}$. Let μ denote the Lebesgue-measure, $R_\alpha : x \to x+\alpha$ (R_α is the rotation by $\alpha 2\pi$). The eigenvalues of R_α are the numbers $e^{2\pi i \{k\alpha\}}$ which implies Kesten's theorem.

We give another example of the relationship between uniform distribution and ergodic theory. This illustrates how the results for distribution of the sequences $(\{n\alpha\})$ imply general results for homeomorphisms of the circle.

Denjoy (1932) proved that for every homeomorphism $T : \mathbb{R}/\mathbb{Z} \to \mathbb{R}/\mathbb{Z}$ having no periodic point there exists an irrational $\alpha(T) \in (0,1)$ such that T is conjugate to the rotation $R_\alpha : x \to x+\alpha$. By this, if we consider (the orbit of a point $x \in \mathbb{R}/\mathbb{Z}$) the distribution of $T^n x$; $n=1,\dots,$ this is determined by the distribution of the sequence $(\{n\alpha\})$. E.g.,

(a) Let $Z_N(I;x)$ denote the number of points $T^n x \in I$, $1 \leq n \leq N$ and μ the invariant measure belonging to T. By Birkhoff's ergodic theorem the remainder term $\Delta_N(I;x) = Z_N(I;x) - N\mu(I) = o(N)$.

By Denjoy's theorem $\Delta_N(I;x)$ is the same as that of the sequence $(\{n\alpha(T)\})$.

(b) The order of points $\{n\alpha\}$, $1 \leq n \leq N$ is very much restricted: if π is the permutation determined by $\{\pi(1)\alpha\} < \dots < \{\pi(N)\alpha\}$, then for example for every α and N fixed $\pi(i) - \pi(i-1)$ takes at most three different values. Now, by Denjoy's theorem the same holds for an arbitrary homeomorphism T and x and permutations π defined by $T^{\pi(1)}(x) < T^{\pi(2)}(x) < \dots < T^{\pi(N)}$. (See T. Sós (1957), Swierczkowski (1958).)

One of the most fascinating and deepest relationships between combinatorics and ergodic theory is given by Furstenberg & al.: Proof and generalization of Szemerédi's theorem. Since there is a recent expository paper by Furstenberg, Katznelson & Ornstein (1982), and the book of Furstenberg (1981), we do not go into the discussion of this.

Strong irregularity

In $[0,1)$ the following "strong irregularity" phenomenon holds.

Theorem 6.5.

A. For every $\varepsilon > 0$ *there exists a* $\delta > 0$ *(depending only on* ε*) such that given an arbitrary sequence* (u_n) *in* $(0,1)$

$$\Delta_n > \delta \log n$$

for all but at most N^ε *values of* $n \leq N$.

B. For every K *there exists a* δ *(depending only on* K*) such that*

$$\Delta_n > K$$

for all but at most $(\log N)^\delta$ *values of* $n \leq N$.

C. For an arbitrary sequence (u_n) *in* $(0,1)$ *the set of values of* x *for which*

$$\Delta_N([0,x)) = o(\log N)$$

holds has Hausdorff dimension 0 .

This theorem was proved first only for $(\{n\alpha\})$ sequences (T. Sós (1979), (1983)), then for arbitrary sequences and in a more general form by Halász (1981) and Tijdeman & Wagner (1980).

One-sided irregularities

Measuring the irregularities with $\|\Delta_N\|_\infty$ or $\|\Delta_N\|_p$, we do not have any information on the sign of the discrepancy. In § 1 we introduced the one-sided discrepancies Δ_N^+ and Δ_N^- . The behaviour of Δ_N^+ and Δ_N^- show some new phenomena.

Again, as almost always, the first results on one-sided irregularities were found for $(\{n\alpha\})$ sequences:

There is no one-sided strong irregularity phenomenon. We mention just the simplest illustrations of this. It is easy to see that

$$\sup_N \Delta_N^+ = \infty \quad , \quad \inf_N \Delta_N^- = \infty \ .$$

However, for an arbitrary sequence $M_N \to \infty$ there exists an α such that

$$\Delta_N^+ < M_N \quad \text{if} \quad N > N_0$$

(resp. there exists an α such that $\Delta_N^- < M_N$).

Similarly, it is easy to see that

$$\Delta_N^+ < K$$

can hold only for a sequence (N_k) of 0 -density. However, for an arbitrary sequence $M_N = o(N)$, there exist an α and a K such that

$$\Delta_n^+ < K$$

holds for at most M_N values of $n \leq N$, if $N > N_0$ (T. Sós (1983)).

Concerning intervals of small discrepancy, first we remark that

$$\sup_N \Delta_N^+([0,\beta)) < \infty$$

may hold also in the case when $\beta \neq \{k\alpha\}$, i.e. when

$$\sup_N \Delta_N([0,\beta)) = \infty .$$

In Dupain & T. Sős (1978) the characterization of the intervals $[0,\beta)$ with

$$\sup_N \Delta_N^+([0,\beta)) < \infty$$

is investigated. Here we mention just one of the new phenomena: there exists an α for which the set $\{\beta | \sup_N \Delta_N^+([0,\beta)) < \infty\}$ is of power of continuum.

However, the assertion in Theorem remains true, if instead of boundedness of $\Delta_N(A)$ we suppose only one-sided boundedness. Halász (1976) proved that if

$$\sup_N \Delta_N^+(A;x) < \infty$$

holds on a set $X \subset \Omega$ of positive measure, then $e^{2\pi i \mu(A)}$ must be an eigenvalue of T.

7 *PARTITION-PROBLEMS FOR GENERAL HYPERGRAPHS*

While in the previous sections we had to make a strong selection because of the large variety and wide scope of the theorems, here we have only limited number of results. Though the problems for general hypergraphs arise from investigations of irregularities on different structures, they are interesting on their own too. To start with, we have to remark that there are almost no lower bounds on the discrepancy of partitions of hypergraphs.

The theorems below give upper bounds for the discrepancy of partitions on hypergraphs. One of the first results of this type is due to Olson & Spencer.

We mention in advance that the results here are related to integer valued programming. The relation between discrepancy of sequences and convex programming was discussed already in Niederreiter (1972).

Theorem 7.1. *(Olson & Spencer (1978)). Let* $\mathcal{A} = \{A_1, \dots, A_k\}$ *be a family of subsets of* S , $|S| = n$. *Consider the two-colorings of* S : $G = \{g \mid g : S \to \{+1, -1\}\}$. *Set*

$$\Delta(\mathcal{A}) = \min_{g \in G} \max_{A \in \mathcal{A}} \left| \sum_{x \in A} g(x) \right|$$

and

$$f(n;k) = \max_{\mathcal{A}} \Delta(\mathcal{A}) ,$$

where the maximum is extended over all families $\mathcal{A}$ *for which* $|\mathcal{A}| = k$ *and* $\bigcup_{A \in \mathcal{A}} A = S$. *Then*

$$f(n;k) < ((n+1)\log 2k)^{1/2} .$$

Theorem 7.2. *(Olson & Spencer (1978)). With the above notation set*

$$h(k) = \max_{\mathcal{A}} \Delta(\mathcal{A})$$

where the maximum is extended over all families $\mathcal{A}$ *for which* $|\mathcal{A}| = k$ *Then*

$$\left(\frac{1}{2} - o(1)\right)k^{1/2} < h(k) < c\, k^{1/2} \log k$$

if a Hadamard matrix of order $k+1$ *exists.*

The upper bound of Theorem 7.2 was improved by Beck & Fiala.

Theorem 7.3. *(Beck & Fiala (1981)).*

$$h(k) < (2k)^{1/2}\sqrt{\log 2k}$$

where c *is a positive absolute constant.*

In most of the partition problems we can easily find a system of linear equations describing the problem and having a fractional solution. How close we can get with integer values to the solution measures how large the discrepancy is.

The theorem of Beck & Fiala below (generalizing a famous result of Baranyai (1975)) is used to get upper bounds for discrepancies.

Theorem 7.4. *(Beck & Fiala (1981)). Integer-making lemma. Let* $\alpha_1,\dots,\alpha_s$ *be given real numbers. Let* A *be a family of subsets of the index set* $\{1,\dots,s\}$ *such that every* $i\in\{1,\dots,s\}$ *belongs to at most* t *members of* A *. Then there exist integers* $a_1,\dots,a_s$ *so that* $|a_i-\alpha_i|<1$, $1\le i\le s$ *and*

$$\left|\sum_{i\in A} a_i - \sum_{i\in A}\alpha_i\right| \le t-1 \quad \textit{for every } A\in A .$$

Theorem 7.5. *(Beck & Fiala (1981)). Let* d *denote the maximal degree of the hypergraph* (S,A) :

$$d(A) = \max_{x\subset S} |\{A\subset A \; ; \; x\subset A\}| .$$

With the notations of Theorem 5.1:

$$\Delta(A) \le 2d(A) - 2 .$$

Conjecture. (Beck & Fiala). For arbitrary (S,A)

$$\Delta(A) \le c(d(A)\log d(A))^{1/2} .$$

Theorem 7.6. *(Beck (1981 a)). With the above notations*

$$\Delta(A) < c(d(A))^{1/2}(\log|A|)^{1/2}\log|S| .$$

Remark. Beck (1981 a) used the above theorems e.g. to prove that Roth's estimate for the discrepancy of arithmetic progressions is nearly sharp. This application itself justifies that the theorem is valuable.

However, it is plausible that upper and lower bounds for

$\Delta(A)$ depending on some structural properties e.g. on some intersection properties of the system A should be found.

We do not know any general theorem for hypergraphs which gives a nontrivial lower bound for the discrepancy $\Delta(A)$.

Olson & Spencer (1978) proved for a system A derived from the Hadamard matrix the discrepancy is greater than $c|A|^{1/2}$. (An analysis of their proof gives a more general lower bound.)

8 *SOME FURTHER PROBLEMS*

We have seen that for different structures different types of discrepancy problems were investigated. Here we give a list of problems for general hypergraphs, implicitly classifying the discrepancy-problems.

Let (S,A) by a hypergraph.

Problem of proportional representation.

Let $\alpha \in (0,1)$ be given. We call an $S_\alpha \subset S$ an α -representative set if

$$(1) \qquad ||S_\alpha \cap A| - \alpha|A|| \leq 1 \quad \text{for every} \quad A \in A .$$

Problem 1.

Find conditions on the existence of an α -representative system in the terms of the structure of the hypergraph.

(1) is a very strong requirement. When it cannot be satisfied, the question is how close can we get to it.

Problem 2.

Find bounds on

$$\min_{S' \subset S} \max_{A \in A} ||S' \cap A| - \alpha|A||$$

(or on

$$\min_{S' \subset S} \Big(\sum_{A \in A} ||S \cap A| - \alpha|A||^p \Big)^{\frac{1}{p}} . \Big)$$

(Does there exist any min-max theorem for some special classes?)

The problems below are formulated for general hypergraphs. Let $\alpha_1,\dots,\alpha_r \in (0,1)$ be given and $\sum \alpha_i = 1$. We say that

$S=\sum_{i=1}^{r} S_{\alpha_i}$, $S_{\alpha_i}\cap S_{\alpha_j}=\emptyset$, $i\neq j$ is an $(\alpha_1,\ldots,\alpha_r)$ -partition

$$||S_{\alpha_i}\cap A| - \alpha_i|A|| \leq 1$$

for every $A\in A$ and $1\leq i\leq r$.

Problem 3.

Find conditions on the existence of an $(\alpha_1,\ldots,\alpha_r)$ -partition.

Let $P(A)\in R^r$ denote the vector $(|S_1\cap A|-\alpha_1|A|,\ldots,|S_r\cap A|-\alpha_r|A|)$.

Problem 4.

Find bounds on

$$\min_{S=\bigcup_{i=1}^{r} S_i} \max_{A\in A} \|P(A)\|$$

where $\|\ \|$ is a norm on R^r (or on

$$\min_{S=\bigcup_{i=1}^{r} S_i} \Big(\sum_{A\in A} ||S\cap A|-\alpha|A||^p\Big)^{\frac{1}{p}}).$$

Problem of well distributed sequences.

Here we consider the "dynamical" version of the previous problem; we want to find a sequence (u_n) in S for which *every* segment is proportionally distributed.

Let (u_n) be a finite resp. infinite sequence in S , and $U_n=\{u_1,\ldots,u_n\}$. We say that (u_n) is uniformly distributed with respect to A if

$$\Delta_n(A) = ||U_n\cap A| - \frac{|A|}{|S|}n| \leq 1$$

for every $n\leq N$ (resp. for every n) and for every $A\subset A$.

Problem 5.

Find conditions on the existence of a uniformly distributed sequence.

Problem 6.

Find bounds on

$$\min_{(u_n)} \max_{n\le N} \max_{A\in \mathcal{A}} \Delta_n(A) .$$

Von Neumann in one of his first papers proved that any everywhere dense sequence in $(0,1)$ can be rearranged to a uniformly distributed sequence.

In the flavor of this simple theorem the following can be asked. We formulate it only in the finite case.

Problem 7.

Let $u_i \in S$, $1\le i\le N$ be a sequence with discrepancy Δ_N. For any permutation π consider the sequence Δ_n^{π}, $n=1,\dots,N$ as the discrepancy of $u_{\pi(1)},\dots,u_{\pi(n)}$. Give bounds on

$$\min_{\pi} \max_{n} \Delta_n^{\pi}$$

or more generally, for a given function $\phi(n)$

$$\min_{\pi} \max_{n} \phi(n)\, \Delta_n^{\pi} .$$

On strong irregularity.

We mentioned in § 6 the "strong irregularity" phenomenon for sequences in $(0,1)$. The analogous question can be asked for many other cases. (In this direction see also Theorem 5.6.)

On one-sided irregularity.

As explained in § 6 new phenomena emerge if we consider the signed discrepancy (Δ_n^+, Δ_n^-) for $(\{n\alpha\})$ sequences. Instead of stating particular questions we just call the attention to this problem.

Decomposition problems.

Problem 1.

The hypergraph (S,A) is called totally unimodular, if every $S'\subset S$ has a partition $g : S' \to \{+1,-1\}$ such that $\left|\sum_{x\in A} g(x)\right| \le 1$ if $A\in \mathcal{A}$.

(This is equivalent to the following: every square submatrix of the incidence matrix of (S,A) has determinant 0 or ± 1.)

Assume there is a partition $A=\bigcup_{i=1}^{d} A_i$ such that (S,A_i), $1\leq i\leq d$ is totally unimodular. Is it true, that there exists a $K=K(d)$ depending only on d such that A has an α-representative system of discrepancy $F(d)$?

Problem 2.

Given Δ and d, does there exist a $\tilde{\Delta}=\tilde{\Delta}(\Delta,d)$ with the following property. If $A=\bigcup_{i=1}^{d} A_i$ and the discrepancy of (S,A_i), $1\leq i\leq d$ is Δ, then the discrepancy of (S,A) is at most $\tilde{\Delta}$.

Added in proof.

Just after having finished this paper, J. Beck proved a surprising and deep theorem. This generalizes several results for families in R^k and gives an answer to the important question, on which geometrical properties does the order of magnitude of the discrepancy depend.

Theorem. *(Beck). Let* S *be a square of sides* N *in* R^2 *and let be given* N^2 *arbitrary points in* S. *Let* B *be a convex domain and* $B(\lambda,\tau,\underline{v})$ *the domain obtained from* B *by a similarity transformation of dilation* $\lambda\leq 1$, *of rotation* τ *and translation by* $\underline{v}\in R^2$. *Let* $\mu(B)$ *resp.* $\ell(B)$ *denote the area resp. the length of the circumference of* B, *and let* $Z(B(\lambda,\tau,\underline{v}))$ *denote the number of points in* $B(\lambda,\tau,\underline{v})$. *Then there exist* λ_0, τ_0, $\underline{v}_0$ *such that*

$$|Z(B(\lambda_0,\tau_0,\underline{v}_0))-\mu(B(\lambda_0,\tau_0,\underline{v}_0)\cap S)| > c\sqrt{\min(\mu(B),\ell(B))}$$

where $c>0$ *is an absolute constant.*

Acknowledgement.

My thanks are due to J. Beck and M. Simonovits for their valuable comments on the paper and to J. Beck for informing me about his very recent results too.

REFERENCES

Ajtai, M., Erdős, P., Komlós, J. & Szemerédi, E. (1981) : On Turán's theorem for sparse graphs. *Combinatorica* 1 (4) 313-317.

Ajtai, M., Komlós, J., Pintz, J., Spencer, J. & Szemerédi, E. (1983): Extremal uncrowded hypergraphs. In manuscript.

Ajtai, M., Komlós, J. & Szemerédi, E. (1981) : A dense infinite Sidon sequence. *European Journal of Combinatorics* 2, 1-11.

Beck, J. (1981 a): Balancing families of integer sequences. *Combinatorica* 1 (3).

Beck, J. (1981 b): Balanced two-colorings of finite sets in the square I. *Combinatorica* 1 (4) 327-335.

Beck, J. (1981 c): Roth's estimate on discrepancy of integer sequences is nearly sharp. *Combinatorica* 1 (4).

Beck, J. (1983 a): Some upper bounds in the theory of irregularities of distribution. *Acta Arithm.* To appear.

Beck, J. (1983 b): Unpublished.

Beck, J. (1983 c): On a problem of K. F. Roth concerning irregularities point distribution. To appear.

Beck, J. (1983 d): Irregularities of two-colorings of the integer coordinate points in the plane. To appear.

Beck, J. & Fiala, I. (1981): Integer making theorems. *Discrete Applied Math.* 3, 1-8.

Beck, J. & Spencer, J. (1983): Well distributed 2-colorings of integers relative to long arithmetic progressions. To appear.

Bollobás, B. (1980): *Extremal Graph Theory*. Academic Press.

Bollobás, B. & Erdős, P. (1976): On a Ramsey-Turán type problem. *J. Comb. Th. Ser. B* 21, 166-168.

Burkill, H. & Mirsky, L. (1973): Monotonicity. *J. Math. Anal. Appl.* 41, 391-410.

Davenport, H. (1956): Note on irregularities of distribution. *Mathematika* 3, 131-135.

Deligne, P. (1975): *Les difféomorphismes du cercle*. Springer, Lecture Notes in Mathematics Séminaire Bourbaki, 477, 01.

Denjoy, A. (1932): Sur les courbes définies par des équations differentielles a la surfacedu tore. *J. de Math. pures et appliquées* 11, 333-375.

Dupain, Y. & T. Sós, V. (1978): On the one-sided boundedness of discrepancy function of the sequence ($\{n\alpha\}$). *Acta Arithm.* 37, 363-374.

Erdös, P. (1964): Problems and results on diophantine approximations. Comp. Math. 16, 52-65.

Erdös, P. (1967): Some recent results on extremal problems in graph theory. *Theory of Graphs*. Gordon & Beach, New York, 117-130.

Erdös, P., Hajnal, A., T. Sós, V. & Szemerédi, E. (1983): On results of Ramsey-Turán type theorems. *Combinatorica*. To appear.

Erdös, P., Meir, A., T. Sós, V. & Turán, P. (1972 b): On some applications of graph-theory III. *Discrete Math.* 2, 207-228.

Erdös, P., Meir, A., T. Sós, V. & Turán, P. (1972 a): On some applications of graph-theory II. *Studies in Pure Math.* Acad. Press, 89-100.

Erdös, P., Meir, A., T. Sós, V. & Turán, P. (1972): On some applications of graph-theory I. *Canadian Math. Bull.* 15 (1), 27-32.

Erdös, P. & Simonovits, M. (1966): A limit theorem in graph theory. *Studia Sci. Math. Hungar.* 1, 51-57.

Erdös, P. & T. Sós, V. (1969): Some remarks on Ramsey's and Turán's theorem. Coll. Math. Soc. Bolyai 4 *Combinatorial Theory*, Balatonfüred, 395-404.

Erdös, P. & T. Sós, V. (1982): On Ramsey-Turán type theorems for hypergraphs *Combinatorica* 2 (3).

Erdös, P. & Spencer, J. (1972): Imbalances in k -colorations. *Networks* 1, 379-385.

Erdös, P. & Stone, M. H. (1946): On the structure of linear graphs. *Bull. Amer. Math. Soc.* 52, 1087-1091.

Erdös, P. & Szemerédi, E. (1972): On a Ramsey-type theorem. *Period. Math. Hungar.* 2, 295-299.

Fürstenberg, H. (1981): *Recurrence in ergodic theory and combinatorial number theory.* Princeton Univ. Press, Princeton, N. J.

Fürstenberg, H., Katznelson, Y. & Ornstein, D. (1982): The ergodic theoretical proof of Szemerédi's theorem. *Bull. Amer. Math. Soc.* 7 (3) 447-654.

Fürstenberg, H., Keynes, H. & Shapiro, L. (1973): Prime flows in topological dynamics. *Israel J. Math.* 14 (1) 26-38.

Gallai, T. (1931): published in Rado (1933 a).

Gerencsér, L. & Gyárfás, A. (1967): On Ramsey-type problems. *Ann. Univ. Sci. Budapest. Eötvös Sect. Math.* 10, 167-170.

Graham, R. L., Rothschild, B. L. & Spencer, J. H. (1980): *Ramsey theory*. Wiley Interscience Series in Discrete Mathematics.

Graham, R. L. (1981): Rudiments of Ramsey theory. Regional conference series in mathematics, No. 45.

Graham, R. L., Leeb. L. & Rotnschild, B. L. (1972): Ramsey's theorem for a class of categories. *Adv. Math.* 8, 417-433.

Halász, G. (1976): Remarks on the remainder in Birkhoff's ergodic theorem. *Acta Math. Hungar.* 27, 389-396.

Halász, G. (1981): On Roth's method in the theory of irregularities of point-distributions. *Proc. Conf. Analytic Number Theory at Durham, 1979.*

Hecke, E. (1922): Über analytische Funktionen und die Verteilung von Zahlen mod Eins. *Abh. Math. Sem. Hamburg* 1, 54-76.

Herman, M. R. (1976): Conjugaison C^{∞} des difféomorphismes du cercle dont le nombre de rotation satisfait a une condition arithmétique. *C. R. Acad. Sci. Paris* 282.

Kesten, H. (1966/67): On a conjecture of Erdős and Szüsz related to uniform distribution mod 1. *Acta Arithm.* 12, 193-212.

Koksma, J. (1942/43): Een algemeene stelling uit de theory der geli kmatige verdeeling modulo 1. *Mathematica B* (Zutphen) 11, 7-11.

Komlós, J., Pintz, J. & Szemerédi, E. (1982): A lower bound for Heilbronn's problem. *Journal of London Math. Soc.*

Kuipers, L. & Niederreiter, H. (1974): *Uniform distribution of sequences. Pure & Applied Mathematics*, Wiley-Interscience.

Niederreiter, H. (1972): On the existence of uniformly distributed sequences in compact spaces. *Compositio Math.* 25, 93-99.

Niederreiter, H. (1978): Quasi-Monte-Carlo methods and pseudo-random numbers. *Bull. Amer. Math. Soc.* 84, 957-1041.

Petersen, K. (1973): On a series of cosecants related to a problem in ergodic theory. *Comp. Math.* 26 (3) 313-317.

Rado, R. (1933 a): Verallgemeinerung eines Satzes von Van der Waerden mit Anwendungen auf ein Problem der Zahlentheorie. Sonderausg. *Sitzungsber. Preuss. Akad. Wiss. Phis. Math. Klasse* 17, 1-10.

Rado, R. (1943): Note on combinatorial analysis. *Proc. London Math. Soc.* 48, 122-160.

Ramsey, F. P. (1930): On a problem of formal logic. *Proc. London Math. Soc.* 30, 264-286.

Roth, K. F. (1952): Sur quelques ensembles d'entiers. *C. R. Acad. Sci. Paris* 234, 388-390.

Roth, K. F. (1954): On irregularities of distribution. *Mathematika* 1, 73-79.

Roth, K. F. (1964): Remark concerning integer sequences. *Acta Arithm.* 9, 257-260.

Roth, K. F. (1979): On irregularities of distribution III. *Acta Arithm.* 35, 373-384.

Roth, K. F. (1980): On irregularities of distribution IV. *Acta Artihm.* 37, 67-75.

Sárközy, A. (1972): See Erdös-Spencer: *Probabilistic Methods in Combinatorics* Akadémiai Kiadó, Budapest 1974.

Schmidt, W. M. (1969): Irregularities of distribution IV. *Inv. Math.* 7 55-82.

Schmidt, W. M. (1972): Irregularities of distribution VII. *Acta Arithm.* 21, 45-50.

Schmidt, W. M. (1974): Irregularities of distribution VIII. *Trans. Amer. Math. Soc.* 198, 1-22.

Schmidt, W. M. (1977): *Lectures on Irregularities of Distribution.* Tata Inst. of Fund. Res. (Bombay), Lectures on Math. and Phys. 56.

Schur, I. (1916): Über die Kongruenz $x^m+y^m \equiv z^m$ mod p. *Jber. Deutsch. Math. Verein.* 25, 114-116.

Simonovits, M. (1968): On methods for solving extremal problems in graph theory stability problems. *Theory of Graphs.* Proc. Coll. held in Tihany, Academic Press, New York, 279-319.

T. Sós, V. (1957): On the theory of diophantine approximation I. *Acta Math. Acad. Sci. Hungar.* 8, 461-472.

T. Sós, V. (1958): On the distribution mod 1 of the sequence $\{n\alpha\}$. *Ann. Univ. Sci. Budapest. Eötvös Sect. Math.* 1, 127-134.

T. Sós, V. (1958): On the theory of diophantine approximation II. *Acta Math. Hungar.* IX (1-2) 229-241.

T. Sós, V. (1968): On a theorem of H. Kesten. *Tagungsbericht Oberwolfach, Zahlentheorie*, p. 18.

T. Sós, V. (1972): On the discrepancy of the sequence $(n\alpha)$. *Tagungsbericht Oberwolfach*, 28.

T. Sós, V. (1974): On the discrepancy of the sequence $(n\alpha)$. *Coll. Math. Soc. J. Bolyai* 13, 359-367.

T. Sós, V. (1969): On extremal problems in graph theory. *Proc. Calgary Internat. Conf. on Comb. Structures*, 407-410.

T. Sós, V. (1979): On strong irregularities of the distribution of $\{n\alpha\}$ sequences. *Tagungsbericht Oberwolfach* 23, 17-18.

T. Sós, V. (1983): Strong irregularities of the distribution of $(n\alpha)$ sequences I. *Studies in Pure Math.* Akadémiai Kiadó. (Turán memorial volume).

T. Sos, V. (1983 b): Strong irregularities and one-sided discrepancy. (Under publication.)

Szemeredi, E. (1973): On graphs containing no complete subgraph with 4 vertices (in Hungarian). *Mat. Lapok* 23, 111-116.

Swierczkowski (1958): On successive settings of an arc on the circumference of a circle. *Fund. Math.* 46, 187-189.

Tijdeman, R. (1980): The chairman assignment problem. *Discrete Math.* 32 323-330.

Tijdeman, R. & Wagner, G. (1980): A sequence has almost nowhere small discrepancy. *Monatshefte Math.* 90, 315-329.

Turån, P. (1941): Eine extremale Aufgabe aus der Graphentheorie (in Hungarian). *Mat. Fiz. Lapok* 48, 436-452. See also: On the theory of graphs. *Colloquium Math.* 3, 19-30.

Van der Corput, J. G. (1935 a): Verteilungsfunktionen I. *Proc. Kon. Ned. Akad. v. Wetensch.* 38, 813-821.

Van der Corput, J. G. (1935 b): Verteilungsfunktionen II. *Proc. Kon. Ned. Akad. v. Wetensch.* 38, 1058-1066.

Van Aardenne Ehrenfest, T. (1945): Proof of the impossibility of a just distribution of an infinite sequence of points over an interval. *Indag. Math.* 7, 71-76.

Weyl, H. (1916): Ober die Gleichverteilung von Zahlen mod Eins. *Math. Ann.* 77, 313-352.

Van der Waerden, B. L. (1927): Beweis einer Bandetschen Vermutung, *Nieuw Arch. Wisk.* 1

ADDENDA AND ERRATA

Certain works cited in the paper have been omitted from the references. They appear to be the following but there was insufficient time in which to confirm this with the author.

Baranyai, Z. (1975): On the factorization of the complete uniform hypergraph. Coll. Math. Soc. Bolyai 10. *Infinite and finite sets*, Keszthely, Vol I, 91-108.

Erdös, P. and Graham, R.L. (1980): Old and new problems and results in combinatorial number theory. *Monographie de l'Enseignement Mathématique*, No. 28.

Hales, A. and Jewett, R.I. (1963): Regularity and positional games. *Trans. Amer. Math. Soc.* 106, 222-229.

Herman, M.R. (1976b): Conjugaison C^∞ des difféomorphismes du cercle pour presque tout nombre de rotation. *C.R. Acad. Sci. Paris* 283.

Olson, J. and Spencer, J.H. (1978): Balancing families of sets. *J. Combin. Theory Ser. A* 25, 29-37.

Rado, R. (1933b): Studien zur Kombinatorik. *Math. Zeit.* 36, 424-480.

Page 219, line 9. For "Erdös-Komlös-..." read "Ajtai-Komlös-..."(?)

E.K.L.

INDEX OF NAMES

This index includes entries for pages on which papers are cited even if the author is not explicitly named there.

SUBJECT INDEX